Math 1100 Workbook

Sam Obeid

Marc Grether

Fall 2006 - Summer 2007

Printed in the United States of America.

ISBN: 0-9741514-5-9

Eagle Images
UNIVERSITY OF NORTH TEXAS
P.O. Box 309615
Denton, TX 76302-9615
940-565-2083

Address all correspondence and order information to the above address.

Introduction

Welcome to Math 1100. We hope that you have a good and instructive experience in this course.

Purpose of this workbook

The purpose of this workbook is to help convey to students the specific material that will be covered in this course. In the past, the authors have noted that many students ask for a detailed list of the topics that they are expected to know to excel on the Common Final. In large part, this workbook is a response to that request.

HOWEVER, *by no means will this workbook (or any other ancillary material) replace the primary importance of class attendance in your success in this course.*

Purpose of the Common Final

The primary purpose of the Common Final for Math 1100 is to ensure that students are held to consistent standards in College Algebra. Each student is expected to know the material detailed in this workbook for the Common Final.

How to use this workbook

Each of Chapters 1-28 cover the sections of material each student is expected to learn in Math 1100.

Each chapter is broken up into **Notes**, **Examples** and **Exercises** sections.

About the Notes

In the **Notes** section, a brief description of some of the material you are expected to master is given. These are intended to reiterate important definitions and describe necessary topics. However, not all necessary definitions are given in this workbook. Other information (particularly context for the definitions) will be developed in class or in other readings.

About the Examples

The **Examples** section provides examples of some common important types of problems, *along with their solutions*. Please note that many times the solution given is not the only possible solution. Your instructor will be able to answer your further questions about this.

Often old final questions are solved in the **Examples** in the appropriate sections. It is our hope that this will help connect the material in each Chapter with the material on the Final.

About the Exercises

The **Exercises** section provides additional problems intended to help students prepare for the Common Final. These problems may be assigned by your instructor. Follow the instructions provided to you by your instructor.

Virtually all of the types of problems on any Common Final are developed in one or more of the sections in this workbook. Reviewing the old Common Finals should reiterate this.

Other Chapter topics

Some chapters contain one or more of the following:

- **Examples** that are started with ✵ are old final problems. Occasionally, different solutions are given in the **Examples** and **Solutions of Old Finals** sections.

- Often it is desirable to check your work. Whenever you see the ✎ symbol, the work involved is checking the answer found to verify its accuracy.

- Many Chapters contain an ❖**Errors to Avoid** section. The errors described here are among those the authors have recognized as common to many students. We point them out in an effort to help you avoid these errors.

About the Old Finals and Solutions of Old Finals

Chapter 29 contains copies of Finals given in Fall 2002, Spring 2003, Spring 2004, Spring 2005, Fall 2005 and Spring 2006. Chapter 30 contains the solutions to the Finals shown in Chapter 29. These are intended to help you check your answers when you work the finals out yourself. Some other old finals are available at http://www.math.unt.edu/ grether/OldFinals/ .

One potential use of the old finals could be to set aside some time to work out each exam starting about one month before the end of the course and progressively working through them (say, doing S03 a month before the final, S04 3 weeks before, F05 2 weeks before, etc.). Once you have completed an old final review the answers and see where you need to concentrate your additional studies.

❖ **Errors to Avoid:** It is best to try to work out the finals yourself and only use the solutions as a last resort or as a means to check yourself. Skipping straight to the solutions doesn't help you determine how much more studying you need to do to learn the material. Also, beginning to work the finals the last week of class is too late. Studying for the Math 1100 Common Final (or any other final) should be ongoing throughout the semester.

Acknowledgements

The revisions contained in this workbook were the product of the hard work and suggestions of many, so it is saddening that some people's contributions will go unacknowledged. Having said that, this workbook would contain many more errors if not for the fastidious eyes and helpful spirits of Rhonda Huettenmueller, Ahmed Rashed, Danielle Bethoney, Paul Ingram, Marcia Edson, Matthew Farmer, Dale Henderson, Hillyer Smith, Richard Ketchersid, Warrawecha Boonsa, Matt Douglass, and Nicholas De La Portilla. The time and energy that you took to try to improve this workbook did not go unnoticed. Thank you.

GRADES

This page is provided to help you keep track of your grades. Please keep all of your assignments until your grade is final.

Assignment Name	Date	Grade		Assignment Name	Date	Grade

Contents

Chapter 1

Graphs of Equations

1.1 Notes about Graphs of Equations

Definition: A *solution* of an equation is a set of values of the variables that make the equation true.

Definition: The *graph of an equation* is the set of all ordered pairs (x, y) that are solutions of the equation.

Definition: There are two types of intercepts on a graph: *x*- and *y*-**intercepts**. x-intercepts are defined as points on a graph that intersect the x-axis. For a point to intersect the x-axis, there must be a solution whose y-value is 0. y-intercepts are defined as points on a graph that intersect the y-axis. For a point to intersect the y-axis, there must be a solution whose x-value is 0.

1.2 ❖ Errors to Avoid

When graphing equations, it is often a good idea to make a table of values to use as a guide. However, be careful in doing so. Each of the three equations $y = x$, $y = x^3$ and $y = 1.5x^4 + x^3 - 1.5x^2$ have the three points $(-1, -1)$, $(0, 0)$, and $(1, 1)$ in common, but their graphs are very different. Often you will need more information about the function than can be determined from the table of values. Later sections will look at many types of graphs in more detail.

1.3   Examples

EX 1. Sketch the graph of $y = 4x - 3$ and find the x- and y-intercepts.

Solution ☞ We will begin by making a table of values. This can be done by selecting a variety of x-values and finding the corresponding y-values. To find some points, including the x-intercept, it may be easier to select the desired y-values and then find the corresponding x-values.

x	y
0	?
?	0
1	?
-1	?
4	?
-4	?

$\Rightarrow$

x	y
0	-3
3/4	0
1	1
-1	-7
4	13
-4	-19

Figure 1.1 below shows each of these points plotted on one graph. Figure 1.2 shows a possible graph of an equation that includes each of the points.

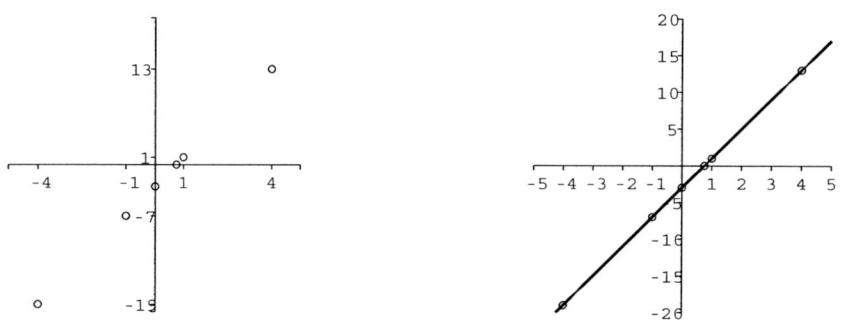

Figure 1.1: Graph of plotted points   Figure 1.2: Graph of $y = 4x - 3$

From the table of values, we see that the x-intercept is located at $\left(\dfrac{3}{4}, 0\right)$ and the y-intercept is located at $(0, -3)$.

EX 2. Sketch the graph of $y^2 = x - 1$ and find the x- and y-intercepts.

Solution ☞ We will begin by making a sign chart and putting these points on a graph.

x	y
0	?
?	0
1	?
?	1
?	2
?	3
?	-1
?	-3

$\Rightarrow$

x	y
0	NS
1	0
1	0
2	1
5	2
10	3
2	-1
10	-3

$\Rightarrow$

Connecting the points we have so far yields Figure 1.3.

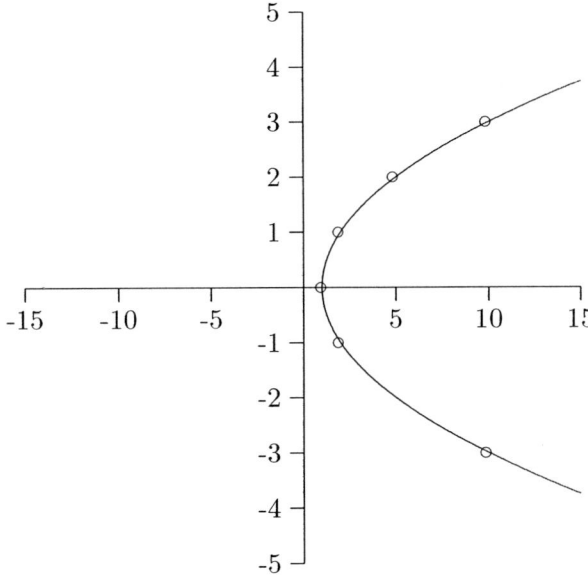

Figure 1.3: Completed graph of $y^2 = x - 1$

From the table of values, we see that the x-intercept is at $(1, 0)$ and there is no y-intercept.

Name:_____

Section Number:_____

Date:_____

Chapter 1 Exercises

For problems 1 - 4, determine whether the given ordered pair is a solution to the given equation and write your answer in the blank provided.

1. $y = 5x - 2$; $(1, \frac{3}{5})$

_____

2. $y^2 = x + 4$; $(12, 4)$

_____

3. $(x - 1)^2 + (y + 2)^2 = 25$; $(1, -2)$

_____

4. $y = (5 - x)^3$; $(6, 1)$

_____

For number 5, fill in the blanks in the chart below for the indicated equation.

5. $y = -3x^2 + 20$

For number 6, sketch the graph of the function used in 5 on the grid provided. Label all intercepts.

6.

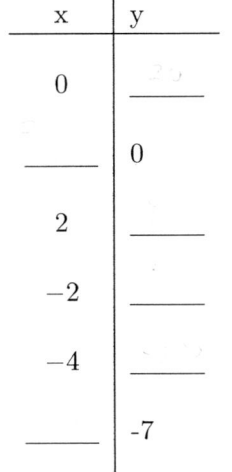

x	y
0	___
___	0
2	___
-2	___
-4	___
___	-7

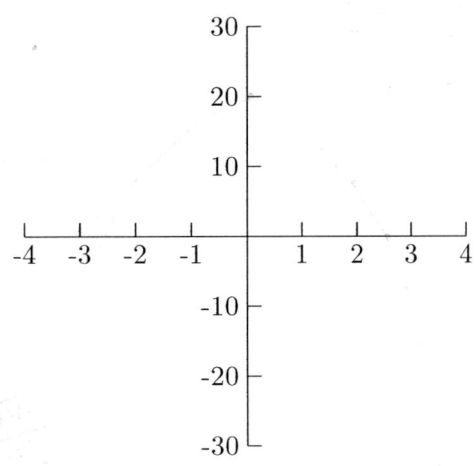

7. Sketch the graph of $(y + 2)^2 = x + 3$ on the graph below. Label the x- and y-intercepts. Show all work.

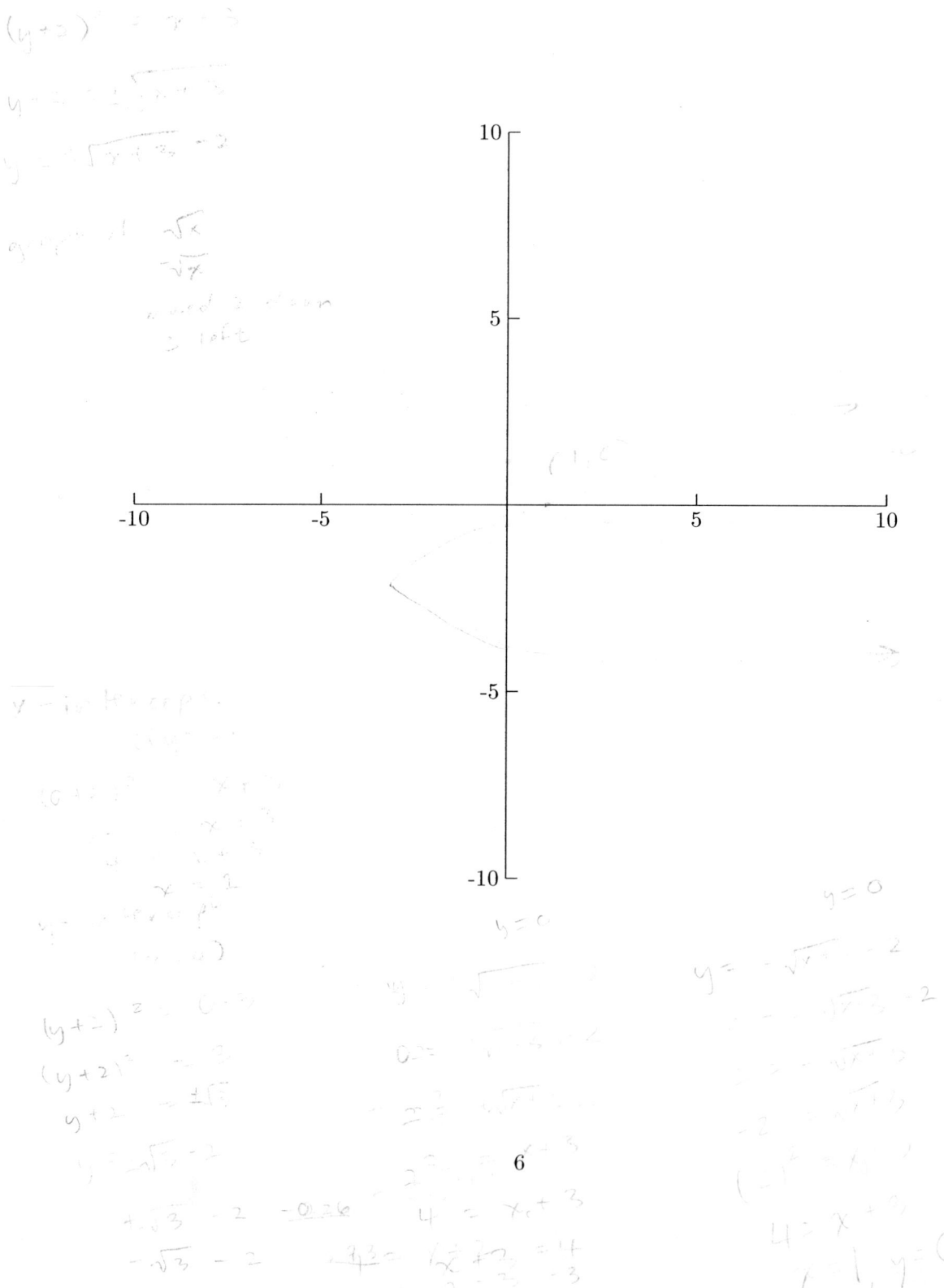

6

For numbers 8 and 9, find the x- and y-intercept(s) from the graphs shown. If necessary, round your answer to one decimal place.

8.

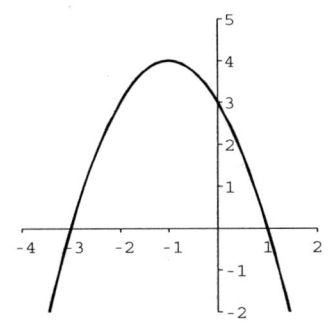

x-intercept(s) _____

y-intercept(s) _____

9.

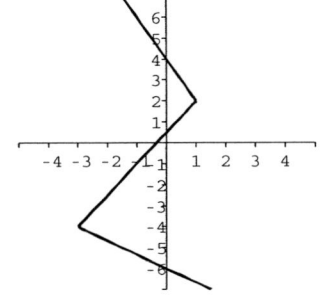

x-intercept(s) _____

y-intercept(s) _____

For problems 10 - 11, determine which of the ordered pairs provided is NOT a solution of the given equation. Show all work and print the answer in the space provided.

10. $3y + 2x = -6$.

  A.) $(-3, 0)$ B.) $(-2, 0)$ C.) $(12, -10)$
  D.) $(3, -4)$ E.) $(0, -2)$

Answer:_____

11. $y = 2x^3 - 8x$.

  A.) $(0, 0)$ B.) $(0, 2)$ C.) $(1, -6)$
  D.) $(2, 0)$ E.) $(-2, 0)$

Answer:_____

7

Chapter 2

Linear Equations

2.1 Notes about Linear Equations

Definition: A *linear equation* in one variable x is an equation that can be written in the form $ax + b = 0$, where a and b are real numbers and $a \neq 0$. To *solve* a linear equation is to find the value of the unknown that make the equation a true statement. Every linear equation has exactly one solution.

Definition: Equations that have the same solution set are called *equivalent equations*. Operations that produce equivalent equations:

 i. Simplifying the equation by removing grouping symbols and combining like terms.

 ii. Adding or Subtracting the same quantities on both sides of the equation.

 iii. Multiplying or dividing by same nonzero number on both sides of the equation.

 iv. Interchanging the two sides of the equation.

To solve a linear equation in x, we need to isolate x on one side of the equation. To do that we can write a sequence of equivalent equations until we reach an equation of the form $x = k$ where k is a real number. This can be done easily if we do the following:

Step 1. Simplify on both sides of the equation (removing grouping symbols and combining like terms).

Step 2. Collect all the terms which involve the variable to one side and the constant terms to the other side of the equation.

Step 3. Divide both sides of the equation by the nonzero coefficient of the variable.

Step 4. Check your answer in the original equation.

2.2 Examples

EX 1. Solve the equation $3x + 12 = 0$ for x .

> **Solution** ☞ Subtracting 12 from both side of the equation $3x + 12 = 0$, we get $3x = -12$. Dividing both sides by 3, we have $x = \dfrac{-12}{3} = -4$. Checking our solution by replacing x with -4 in the original equation. $3(-4) + 12 = -12 + 12 = 0$, hence our solution satisfies the equation.

EX 2. Solve the equation $3(x - 2) + x = -10 - 2(x - 4)$.

> **Solution** ☞ We start by simplifying the algebraic expression on each side of the equation.
> $$3(x - 2) + x = -10 - 2(x - 4) \Rightarrow$$
> $$3x - 6 + x = -10 - 2x + 8 \Rightarrow$$
> $$4x - 6 = -2 - 2x$$
>
> Now collect all the terms which involve the variable x to one side and the constant terms to the other side of the equation
> $$4x + 2x = -2 + 6 \Rightarrow 6x = 4$$
>
> To isolate x, we divide both sides of the equation by 6,
> $$x = \frac{4}{6} \Rightarrow x = \frac{2 \cdot 2}{3 \cdot 2} \Rightarrow x = \frac{2}{3} \cdot 1 = \frac{2}{3}$$
>
> In the next three examples the equations involve fractions. A good strategy in such case is eliminate the fractions by multiplying both sides of the equation by the least common denominator.

EX 3. Solve the equation $\dfrac{3x}{2} - \dfrac{1}{3} = \dfrac{x}{3} - 5$.

> **Solution** ☞ The least common denominator of 2, 3, and 3 is 6. Multiplying both sides of the equation $\dfrac{3x}{2} - \dfrac{1}{3} = \dfrac{x}{3} - 5$ by 6 yields
> $$9x - 2 = 2x - 30$$
>
> Collecting the terms which involve x to one side and the constant terms to the other side of the equation gives
> $$9x - 2x = -30 + 2 \Rightarrow 7x = -28$$
>
> Dividing both sides of the equation by 7 leaves us with
> $$x = \frac{-28}{7} \Rightarrow x = -4.$$

EX 4. Solve the equation $\dfrac{4}{x} + \dfrac{1}{2} = \dfrac{3}{5}$.

Solution ☞ The least common denominator of x, 2, and 5 is $10x$. Multiplying both sides of the equation $\dfrac{4}{x} + \dfrac{1}{2} = \dfrac{3}{5}$ by $10x$ yields

$$40 + 5x = 6x \Rightarrow 40 = 6x - 5x \Rightarrow 40 = x \text{ or } x = 40.$$

Here we must check if our solution $x = 40$ makes any of the denominators zero. If it does we have to reject the solution (see the next example). In this case it does not, so $x = 40$ is the solution.

EX 5. Solve the equation $\dfrac{3}{x-2} - \dfrac{1}{x+2} = \dfrac{12}{(x+2)(x-2)}$.

Solution ☞ Here the least common denominator of $x - 2$, $x + 2$, and $(x+2)(x-2)$ is $(x-2)(x+2)$. If we multiply both sides of the equation $\dfrac{3}{x-2} - \dfrac{1}{x+2} = \dfrac{12}{x^2-4}$ by $(x-2)(x+2)$, then we have

$$3(x+2) - (x-2) = 12 \Rightarrow 3x + 6 - x + 2 = 12 \Rightarrow$$

$$2x + 8 = 12 \Rightarrow 2x = 4 \Rightarrow x = 2$$

✐ Checking to see if $x = 2$ is a solution, we see that $x = 2$ makes the first denominator in the equation zero. Therefore this equation has no solution.

Name:_____

Section Number:_____

Date:___07/11/2022_____

Chapter 2 Exercises

Solve the equations given in numbers 1 - 5.

1. $3x - 10 = 11$

   $3x = 11 + 10$

   $3x = 21$

   $x = \frac{21}{3}$

   $x = 7$

Verification, $x = 7$

$3(7) - 10 = 11$

$21 - 10 = 11$

$11 = 11$

$LS = RS$

2. $2(x - 7) - 5x = 17 - 6x$

   $2x - 14 - 5x = 17 - 6x$

   $2x - 5x + 6x = 17 + 14$

   $3x = 31$

   $x = \frac{31}{3}, \ 10.\overline{3}$

verification, $x = \frac{31}{3}, \ 10.\overline{3}$

$2(\frac{31}{3} - 7) - 5(\frac{31}{3}) = 17 - 6(\frac{31}{3})$

$2(3.\overline{3}) - 5(10.\overline{3}) = 17 - 6(10.\overline{3})$

$6.\overline{6} - 51.\overline{6} = 17 - 62$

$-45 = -45$

3. $4 - 5(x - 3) + 8x = 15 + 6(1 - 2x)$

   $4 - 5x + 15 + 8x = 15 + 6 - 12x$

   $-5x + 8x + 12x = 15 + 6 - 4 - 15$

   $15x = 2$

   $x = \frac{2}{15}$

verification, $x = \frac{2}{15}$,

$4 - 5[(\frac{2}{15}) - 3] + 8(\frac{2}{15}) = 15 + 6[1 - 2(\frac{2}{15})]$

4. $y - 2[3 - (y + 9)] = 12 - (y - 7) \quad -y \ -(-7)$

   $y - 2[3 - y - 9] = 12 - y + 7 \rightarrow y = \frac{7}{4}, 1\frac{3}{4}, 1.75$

   $y - 6 + 2y + 18 = 19 - y$

   $3y + 12 = -y + 19$

   $4y = 7$

5. $\dfrac{x - 1}{2} - \dfrac{3x - 2}{4} = \dfrac{7 - 2x}{6}$

   $\dfrac{x-1}{2} \cdot \dfrac{12}{12} - \dfrac{3x-2}{4} \cdot \dfrac{6}{6} = \dfrac{7-2x}{6} \cdot \dfrac{4}{4}$

   $= \dfrac{12(x-1)}{24} - \dfrac{(3x-2)(6)}{24} = \dfrac{14(7-2x)}{24}$

   $24\left[\dfrac{12(x-1) - 6(3x-2)}{24}\right] = \left[\dfrac{4(7-2x)}{24}\right] \cdot 24$

   $12x - 12 - 18x + 12 = 28 - 8x$

   $-6x = 28 - 8x$

   $2x = 28 \quad x = \dfrac{28}{2} \quad x = 14$

$5 \cdot \frac{1}{3} = \frac{5}{3}$

$1\frac{2}{3}$

$6 \cdot \frac{1}{3} =$

$\dfrac{a}{b} + \dfrac{c}{b} = \dfrac{d}{b} \ \}$ simplify

$a + c = d$

$\not{b}\left(\dfrac{a+c}{\not{b}}\right) \ \not{b}\left(\dfrac{d}{\not{b}}\right)$

$$\frac{2}{x} - \frac{3}{2x} = \frac{1}{8}$$

$$\frac{2(8)}{x(8)} - \frac{3(4)}{2x(4)} = \frac{1(x)}{8(x)}$$

$$8x\left(\frac{16-12}{8x}\right) = 8x\left(\frac{x}{8x}\right)$$

$$\frac{16}{8x} - \frac{12}{8x} = \frac{x}{8x}$$

$$4 = x$$

$$x = 4$$

Solve the equations given in numbers 6 - 9.

NPV

$x \neq 0$

6. $\dfrac{2}{x} - \dfrac{3}{2x} = \dfrac{1}{8}$

$$\frac{2(2)}{x(2)} - \frac{3}{2x} = \frac{1}{8}$$

$$\frac{4-3}{2x} = \frac{1}{8}$$

$$\frac{1}{2x} = \frac{1}{8}$$

$$1 = 2x \cdot \frac{1}{8}$$

$$1 = \frac{2x}{8}$$

$$1 = \frac{x}{4}$$

$$4 = x$$

$$x = 4$$

verification.

$$\frac{2}{4} - \frac{3}{2(4)} = \frac{1}{8}$$

$$\frac{4}{8} - \frac{4}{8} = \frac{1}{8}$$

7. $\dfrac{3}{x+3} - \dfrac{4}{x-2} = \dfrac{5}{2x+6}$

NPV

$x \neq 2, -3$

$$\frac{3}{x+3} - \frac{4}{x-2} = \frac{5}{2(x+3)}$$

$$\frac{3(x-2)}{(x+3)(x-2)} - \frac{4(x+3)}{(x-2)(x+3)}$$

$$\frac{3(x-2)-4(x+3)}{(x+3)(x-2)} = \frac{5}{2(x+3)}$$

$$3(x-2) - 4(x+3) = \frac{5(x+3)(x-2)}{2(x+3)}$$

$$2\left[3(x-2) - 4(x+3)\right] = 5(x-2)$$

$$6(x-2) - 4(x+3) = 5x - 10$$

$$6x - 12 - 4x - 12 = 5x - 10$$

$$6x - 4x - 5x = +24 - 10$$

$$-3x = 14$$

$$x = -\frac{14}{3}, -4\frac{2}{3}, -4.\overline{6}$$

8. $\dfrac{6}{x-2} - \dfrac{1}{x+5} = \dfrac{2}{x^2+3x-10}$

NPVs

$x \neq 2, -5$

$$\frac{6}{x-2} - \frac{1}{x+5} = \frac{2}{(x+5)(x-2)}$$

$$\frac{6(x+5)}{(x-2)(x+5)} - \frac{1(x-2)}{(x+5)(x-2)} = \frac{2}{(x+5)(x-2)}$$

$$6(x+5) - (x-2) = 2$$

$$6x + 30 - x + 2 = 2$$

$$6x - x = -30$$

$$5x = -30$$

$$x = -30/5$$

$$x = -6$$

9. $\dfrac{x}{x-1} + 3 = \dfrac{1}{x-1}$

NPVs

$x = 1$

$$\frac{x}{x-1} - \frac{1}{x-1} = -3$$

$$\frac{(x-1)}{(x-1)} = -3$$

$$1 \neq -3$$

no real solutions

alt

$$\frac{x}{x-1} + \frac{3(x-1)}{(x-1)} = \frac{1}{x-1}$$

$$\begin{array}{r}36 \\ -18 \\ -10 \\ \hline 8\end{array}$$

$$\frac{-6\,\tiny{2/3}}{}$$

$$\frac{1}{-1}$$

14

10. The length of a rectangular parking lot is three times its width. If the perimeter is 1,200 meters, find the lot's dimensions.

150m by 750m

x width

3x
length

$3x + x + 3x + x$

$2(4x)$ total perimeter =

$2(3x + x)$ $8x$

11. A rectangular parcel of land is 5ft longer than it is wide and has a perimeter of 90ft. Find the dimensions of this lot.

x

$x + 5$

$x + 5 + x + 5 + x + x = 90$

$4x + 10 = 90m$

$4x = 80m$

$x = 20m$

25m by 20m

For numbers 12 - 16, show all work and print the answer in the space provided.

12. Solve the equation $2x + 23 = 7(x - 1)$ for x.

A.) $x = \dfrac{24}{5}$ B.) $x = 4$ C.) $x = 5$

D.) $x = 7$ E.) None of these.

$2(6) + 23 = 7(6-1)$

$12 + 23 = 7(5)$

$35 = 35$

$LS = RS$

$2x + 23 = 7x - 7$

$2x - 7x = -7 - 23$

$-5x = -30$

$x = -30 / -5$

$x = 6$

Answer:_____

13. Solve the equation $\dfrac{3}{2x} - 5 = -\dfrac{1}{x}$ for x.

A.) $x = 2$ B.) $x = 5$ C.) $x = -\dfrac{1}{2}$

D.) $x = \dfrac{1}{2}$ E.) None of these.

$\dfrac{3}{2x} - 5 \cdot \dfrac{2x}{2x} = -\dfrac{1}{x} \cdot \dfrac{2}{2}$

$\dfrac{3}{2x} - \dfrac{10x}{2x} = -\dfrac{2}{2x}$

$2x\left[\dfrac{3 - 10x}{2x}\right] = \left[\dfrac{-2}{2x}\right] \cdot 2x$ 15

$3 - 10x = -2$

Answer:_____

$-10x = -2 - 3$

$-10x = -5$

$x = \dfrac{-5}{-10}$

$x = \dfrac{1}{2}, 0.5$

14. Solve the equation $3(t - 1) - 7(t + 2) = 1$ for t.

A.) $t = -\dfrac{1}{2}$ B.) $t = -\dfrac{9}{2}$ C.) $t = \dfrac{5}{2}$

D.) $t = 0$ E.) None of these.

$3t - 3 - 7t - 14 = 1$

$3t - 7t = 1 + 3 + 14$

$-4t = 18$

$t = -\dfrac{18}{4}$ $t = -\dfrac{9}{2}$

Answer:_____

15. Solve the equation $\dfrac{2}{x - 3} - \dfrac{3}{x + 3} = \dfrac{10}{x^2 - 9}$ for x.

A.) $x = \dfrac{1}{2}$ B.) $x = -1$ C.) $x = 5$

D.) $x = 9$ E.) None of these.

$\dfrac{2(x+3)}{(x-3)(x+3)} - \dfrac{3(x-3)}{(x+3)(x-3)} = \dfrac{10}{(x+3)(x-3)}$

$1 - \dfrac{3}{8} = \dfrac{10}{2x-9}$

$1 - \dfrac{3}{8} = \dfrac{10 \div 2 = 5}{16 \div 2 = 8}$

$\dfrac{5}{8} = \dfrac{5}{8}$

$2x + 6 - 3x + 9 = 10$

$-x + 15 = 10$

$-x = -5$ $x = 5$

Answer:_____

16. Solve the equation $\dfrac{t}{t - 2} - 4 = \dfrac{2}{t - 2}$ for t.

A.) $t = \dfrac{1}{2}$ B.) $x = -2$ C.) $t = 2$

D.) $t = 6$ E.) None of these.

$\dfrac{t}{t-2} - \dfrac{2}{t-2} = 4$

$\dfrac{t-2}{t-2} = 4$

$1 \neq 4$

Answer:_____

Chapter 3

Applied Problems in One Variable

3.1 Notes about Applied Problems in One Variable

A good method to use in solving Applied Problems is George Polya's four-step process. The four steps are:

- *Understanding the problem.* To fully understand the problem it is sometimes necessary to reread the question, draw a diagram, or to do a related, yet simpler problem. It is also useful to give variable names to important criteria in the problem.

- *Developing a plan.* You will need to determine how you are going to go about solving the problem. For the purposes of this class, you should expect to set up an algebraic equation that you can help you solve the problem. Spend time evaluating whether your equations is properly constructed.

- *Following the plan.* Once you have determined the equation(s), solve them for the appropriate variables. If you find that you do not have information in the form of the equation(s) to solve the problem, reread the problem to see if there was any information that was not used. If so, try to find other equations.

- *Looking back.* This step includes such steps as checking your answer and determining whether the answer you got seems "reasonable". By checking for reasonableness, you may find that though you performed the algebra correctly, the answer is noticeably wrong because you set up the equation incorrectly.

Though you may find that you are able to do some problems without using these steps, it is a very good idea to be familiar with them. When you come to a problem that you are unable to do, these steps provide a basis for how to go about solving the problem, step by step. They give you the freedom to try different options, but also the power to know when you need to go in a different direction.

3.2   ❖ Errors to Avoid

If you are setting up equations to solve, be sure to equate *equal* amounts as well as equal *types* of things. For example, distances should be equated to distances, rates to rates, interest to interest, etc. Make sure to have a good understanding of the problem and make sure that you carefully define your variables.

Be careful to consider whether you have used all of the given information. If you have not, be sure that the 'extra' information was extraneous.

3.3   Examples

EX 1. A high speed train travels for 6 hours at a speed of 120 mph. What length of time (in hours) could be cut off of the trip if the train went 200 mph instead?

Solution ☞ *Understanding the Problem*: Here we will need to make use of the relationship $Distance = Rate \cdot Time$ or in its abbreviated form $D = R \cdot T$. Also note that to determine the length of time cut off, we only need to determine the length of time that the trip takes at 200 mph and then subtract 6 hrs, so we'll let x be the length of time that the train would need to travel at 200 mph.

Developing a plan: As noted above, we are looking for a time. Because of this, we will need to know both a distance and a rate to use the equation $D = R \cdot T$. We are given that the train is traveling 200 mph, but we need to determine the distance the train must travel. In this case, the train must travel $D = 6$ hours $\cdot 120 \dfrac{\text{miles}}{\text{hour}}$ or 720 miles. Thus, to find the amount of time the train would travel at 200 mph, we need only solve $720 = 200 \cdot T$.

Following the plan: Solving 720 miles $= 200 \dfrac{\text{miles}}{\text{hour}} \cdot T$ by dividing both sides by 200 yields $T = 3.6$ hours.

Looking back: If a train travels 3.6 hours at 200 mph, it will travel 720 miles. A train traveling 6 hours at 120 mph will also travel 720 miles, so our solution does work.

EX 2. A recipe calls for 3 cups of 35% strength vinegar, but Maggie only has bottles with 40% and 20% strength vinegar. How much of the 20% vinegar should Maggie mix to make 3 cups of 35% vinegar?

Solution ☞ *Understanding the Problem*: Maggie will be using both types of vinegar, but you are only asked about how much of the 20% vinegar will be used. Let x be the amount of 20% vinegar used. Thus, 3-x is the amount of 40% vinegar to be used.

Developing a plan: We need to find an equation that relates two things that are *equal* for this problem.

It seems that something along the lines of $.2(x) + .4(3 - x) = 3$ should be what we want.

However, the $.2(x)$ and $.4(3-x)$ each represent the amount of vinegar in the solutions, while the 3 represents the *total* amount of solution, not just the vinegar in the solution. Therefore, these amounts do not represent equal amounts.

✎ It seems that the initial equation was on the right track, but the RHS and LHS represent different types of things. Thankfully this is easy to correct. Rather than using 3 as the RHS, $.35(3)$ should be used because this represents the amount of vinegar in the final solution.

Following the plan: For this problem we must solve the equation: $.2(x) + .4(3-x) = .35(3) \Rightarrow .2x + 1.2 - .4x = 1.05 \Rightarrow -.2x = -.15 \Rightarrow x = .75$ cups.

Our solution is that Maggie will use .75 cups of the 20% vinegar solution.

Looking back: Using .75 cups of the 20% vinegar solution means that 2.25 cups of the 40% vinegar solution would be used. Because 35% is closer to 40% than to 20%, we would expect that more of the 40% solution would be needed. This doesn't prove that the answer is correct, but it should make us feel more confident.

✎ A more thorough check shows that both the LHS and RHS are equal to 1.05. We see that $.2(.75) + .4(3 - .75) = .15 + .9 = 1.05$ and also that $.35(3) = 1.05$.

EX 3. ⋆ (S04#16) Ahmed plans to invest his $8,000 into two different bank accounts for one year. One account pays 6% annual simple interest and the other account pays 3% annual simple interest. Ahmed wants to earn a total of $375 in interest in one year. How much money should he invest into the account that pays 6% interest?

Solution ☞ *Understanding the Problem*: Ahmed will be putting money into two separate accounts and the total amount of money is $8,000. If x represents the amount of money going into the account earning 6%, then $8,000 - x$ remains. Thus, $8,000 - x$ must be the amount being deposited into the account earning 3%.

Developing a plan: Our plan is to determine an equation involving x that relates to the interest Ahmed wants to earn. Because it is known that Ahmed wants to earn a total of $375 in interest, it seems that the amount of interest earned on each account should play a role. The amount of interest earned on the account earning 6% interest is $.06(x)$ and the amount of interest earned on the account earning 3% interest is $.03(8000 - x)$. Thus, the total amount of interest earned is $.06(x) + .03(8000 - x)$.

Following the plan: We've found that the total amount of interest is given by $.06(x) + .03(8000 - x)$ and should equal $375. Thus, we need to solve $.06(x) + .03(8000 - x) = 375$. Distributing throughout yields $.06x + 240 - .03x = 375$. Combining like terms on the LHS and subtracting 240 from each side gives $.03x = 135$. Dividing both sides by $.03$ gives $x = 4500$.

Looking back: If Ahmed has $4,500 in the account that earns 6%, he will earn $270 in interest. The remaining $3,500 would earn $105 for a total of $375 in interest. Thus, $4,500 should be deposited into the account paying 6%.

EX 4. It takes Apu 3 hours to stock the shelves and it takes Homer 7 hours to stock the shelves. How long will it take them to stock the shelves if they work together?

Solution ☞ *Understanding the Problem*: Here we are looking for an amount of time, so we will call x the amount of time in hours it takes Apu and Homer to complete the job. It's also important to note that it should take them less time if they are working together than if either were doing the job alone.

Developing a plan: To solve this problem, we will need to develop a relationship between equal quantities. As noted earlier, these quantities should have the same units. Looking at the information given, we see that we are given the amount of time it takes each of Apu and Homer to complete the task. However, it isn't clear how that relates to the overall time. In particular, subtracting $7 - 3$ yields 4, which can't be correct.

One way to use the information given would be to consider the rate at which Apu and Homer completed 1 job each. Further, this rate multiplied by the time (which should be x for each) would represent the part of the job each has completed. Because Apu and Homer are working together, we can add these together to yield one completed job.

Following the plan: First, we'll find the rate at which Apu and Homer complete one job. Apu can complete 1 job in 3 hours, so he completes $1/3 \, \frac{\text{job}}{\text{hour}}$. Homer can complete 1 job in 7 hours or $1/7 \, \frac{\text{job}}{\text{hour}}$. Thus, in x hours Apu and Homer complete $\frac{x}{3}$ and $\frac{x}{7}$ of the job each or $\frac{x}{3} + \frac{x}{7}$ together. Because they are trying to complete 1 job, we'll get $\frac{x}{3} + \frac{x}{7} = 1$. To solve this, we'll multiply each term by 21 and get $7x + 3x = 21$. Combining the like terms and dividing both sides by 10 gives that $x = 21/10$ or 2.1 hours.

Looking back: In 2.1 hours, Apu would complete $\frac{1}{3} \cdot 2.1$ or .7 of a job and Homer would complete $\frac{1}{7} \cdot 2.1$ or .3 of the job. Together, they'd have completed one job.

EX 5. An item is marked for sale at 20% off the regular price. The item does not sell and is then marked down an additional 40% off the already discounted price. The item now sells for $ 25.92. Find the original price.

Solution ☞ *Understanding the Problem*: Here we are looking for the original price, so we will let x represent the original price of the item. Note that marking an item 20% off and then taking an additional 40% off the discounted price *is not* the same as marking 60% off the original price.

Developing a plan: We will try to determine a formula to relate x to the original discount price. The original discounted price would be $x - .2x$ or $.8x$. So the reduced discount price would be $.8x - .4(.8x)$ or $.48x$.

Following the plan: Solving $.48x = 25.92$ gives $x = 54$.

Looking back: If the original price is $54, then the price after a 20% discount would be $43.20. If the item is marked down 40% off of the $43.20, then the final price will be $25.92.

Chapter 3 Exercises

Solve problems 1 - 7. Be sure to show all work.

1. Sarah sells a fruit juice that she makes by mixing together fruit juice concentrate and water. She wants her mixture to contain 30 liters of 15% strength juice. How much of the fruit juice concentrate will she need if the concentrate contains 90% juice?

I have no idea what's being asked here

2. Train *A* heads east in a straight line from Gausston towards Euler city at a constant 30 mph. Train *B* heads west from Euler city towards Gausston at 40 mph. If the trains each leave at the same time and they pass each other 1.5 hours later, how far apart are Gausston and Euler City?

Train A 30mph   Gausston   60 miles   Euler City   Train B 40mph   45 miles

miles per hour

$$\frac{30\ miles}{hour} \cdot 1.3h =$$

45 miles

$$\frac{40\ miles}{hour} * 1.5h$$

60 miles

105 miles

23

3. Rachel is investing money in two different funds for one year, Joey Mutual Fund and Ross Bond Fund. The Joey fund is expected to yield 12% simple interest and the Ross Fund is expected to yield 4% simple interest. If Rachel invests a total of $6,000 into the two funds and she wants to receive $440 in interest during the year, how much money should she put in the Ross Fund?

$$0.04(x) + 0.12(6000 - x) = 400$$

4. It takes Tom 6 hours to paint a fence and Huck 4 hours to paint a fence. If they work together, how long will it take them to paint the fence?

Tom

1 job, in 6 hours

$$\frac{1 \text{ job}}{6 \text{ hours}}$$

$$\frac{1}{6} \frac{\text{job}}{\text{hour}},$$

Huck

$$\frac{1 \text{ job}}{4 \text{ hours}}$$

$$\frac{1}{4} \frac{\text{job}}{\text{hour}}$$

$$10x = 24$$

$$x = 2.4 \text{ h}$$

$$\frac{x}{6} + \frac{x}{4} = 1$$

$$\frac{4x}{24} + \frac{6x}{24} = 1$$

5. Jack is walking slowly away from a lamp post. After a short time, he notices that the tip of his shadow reaches to the outer edge of the light's ray. At this point, he notices that he is 8 ft away from the tip of his shadow and 32 ft from the lamp post. If Jack is 6 ft tall, how high is the lamp post?

I also have no idea what is being asked here

proportion?

$$\frac{8ft}{24ft} = \frac{6ft}{x}$$

18 ft (?)

6. Frank the farmer is building a compost bin. He wants the bin to hold 5 cubic meters of material. If the bin is a right circular cylinder and has a base with a radius of 90 cm, how high should the bin be (round your answer to the nearest cm)?

0.9m

$$\frac{5 m^3}{\pi \cdot (0.9m)^2}$$

=

area of a circle = πr^2

0.9 m

7. A sprint triathlon is a race that consists of, in order, a 0.75 km swim, a 20 km bike and then a 5 km run. Both Olympic Bob and Average Joe are competing in the race. Bob can swim 3 kilometers per hour (or kph), bike 30 kph, and run 15 kph. Olympic Bob is 15 minutes late for the race. Joe can swim 2.25 kph, bike 25 kph and run 10 kph.

(a) Who is ahead after the swimming leg of the race? What distance is the leader ahead by when he finishes swimming?

$$OB: \quad 0.25h + 0.75km \cdot \frac{1h}{3\ km}$$

$$AJ: \quad 0.75km \cdot \frac{1h}{2.25km}$$

(b) Who is ahead after the biking leg of the race? What distance is the leader ahead by when he finishes biking?

(c) Who is ahead after the running leg of the race? How much longer (in seconds) will it take the loser to finish the race?

Chapter 4

Complex Numbers

4.1 Notes about Complex Numbers

Definition: A *Complex Number* is an expression of the form $a + bi$, where a and b are real numbers and $i^2 = -1$. a is called the real part of $a + bi$ and b is the imaginary part of $a + bi$.

Adding, Subtracting, and Multiplying Complex Numbers:

i. $(a + bi) + (c + di) = (a + c) + (b + d)i$. To add complex numbers, add the real parts and the imaginary parts.

ii. $(a + bi) - (c + di) = (a - c) + (b - d)i$. To subtract complex numbers, subtract the real and imaginary parts.

iii. $(a + bi)(c + di) = (ac - bd) + (ad + bc)i$. To multiply two complex numbers, use the FOIL method than replace i^2 with -1 and simplify.

Equality of Complex Numbers: Two complex numbers are equal if and only if their real parts are equal and their imaginary parts are equal. $a + bi = c + di \Leftrightarrow a = c$ and $b = d$.

Conjugate of a Complex Numbers: The conjugate of a complex number $z = a + bi$ is $\overline{z} = a - bi$. The product of a complex number $z = a + bi$ and its conjugate $\overline{z} = a - bi$ is always a positive real number given by $z \cdot \overline{z} = (a + bi)(a - bi) = a^2 + b^2$.

Dividing Complex Numbers: Dividing a two complex numbers is equivalent to simplifying the quotient $\dfrac{a+bi}{c+di}$. To do this, multiply the numerator and the denominator by the conjugate of the denominator, which will give:
$$\frac{a+bi}{c+di}=\frac{a+bi}{c+di}\cdot\frac{c-di}{c-di}=\frac{(ac+bd)+(bc-ad)i}{c^2+d^2}.$$

Powers of i: Since $i^4=1, i^{4k}=(i^4)^k=1^k=1$ for all integer k. So i raised to the power of any multiple of 4 is 1. Any positive integer n can be written in the form $n=4k+r$ where $r=0,1,2$ or 3 is the remainder of the division of n by 4. Therefore $i^n=i^{4k+r}=i^{4k}i^r=1i^r=i^r$. Hence, $i^n=i^r$ where r is the remainder of the division of n by 4.

1. If $r=0$, then $i^n=1$

2. If $r=1$, then $i^n=i^1=i$

3. If $r=2$, then $i^n=i^2=-1$

4. If $r=3$, then $i^n=i^3=-i$

Principle Square Roots of Negative Numbers: If $-r$ is a negative number, then the Principal square root of $-r$ is $\sqrt{-r}=i\sqrt{r}$.

Imaginary Roots of Quadratic equations: Recall that the solution of a quadratic equation $ax^2+bx+c=0$ is $x=\dfrac{-b\pm\sqrt{b^2-4ac}}{2a}$. If the discriminant $b^2-4ac<0$, then the equation has no real solution. But in the complex number system this equation will have solutions because negative numbers have square roots in the set of complex numbers.

4.2 ❖ Errors to Avoid

$\sqrt{a}\cdot\sqrt{b}=\sqrt{ab}$ only when at least one of a and b are nonnegative real numbers. In general to correctly solve problems of this type, first determine the principle square root of each number and then multiply the numbers (see Example 8).

Be careful when dividing terms such as $\dfrac{-4\pm2i}{2}$. The solution is not $-2\pm2i$. For the correct solution, see Example 10.

4.3   Examples

EX 1. What are the real and complex parts of the complex number $2 - \sqrt{3}i$?

> **Solution** ☞ The real part of $2 - \sqrt{3}i$ is 2 and the imaginary part is $-\sqrt{3}$.

EX 2. Write $(3 + 5i) - (4 - 2i)$ in the form $a + bi$.

> **Solution** ☞ $(3 + 5i) - (4 - 2i) = 3 + 5i - 4 + 2i = -1 + 7i$.

EX 3. Write $(-3i) \cdot (2 - 7i)$ in the form $a + bi$.

> **Solution** ☞ $(3i)(2 - 7i) = -6i + 21i^2 = -21 - 6i$.

EX 4. Write $(3 + 5i) \cdot (4 - 2i)$ in the form $a + bi$.

> **Solution** ☞ We start by using the FOIL method.
>
> $(3 + 5i) \cdot (4 - 2i) = 12 - 6i + 20i - 10i^2 = 12 + 14i + 10 = 22 + 14i$.

EX 5. Write $(\frac{1}{2} + i\frac{\sqrt{3}}{2})(\frac{1}{2} - i\frac{\sqrt{3}}{2})$ in the form $a + bi$.

> **Solution** ☞ Again if we use the FOIL method we get
>
> $(\frac{1}{2} + i\frac{\sqrt{3}}{2})(\frac{1}{2} - i\frac{\sqrt{3}}{2}) = \frac{1}{4} - \frac{\sqrt{3}}{4}i + \frac{\sqrt{3}}{4}i - \frac{\sqrt{9}}{4}i^2 = \frac{1}{4} + \frac{3}{4} = 1$.

EX 6. Write $\dfrac{2 + 3i}{1 - 2i}$ in the form $a + bi$.

> **Solution** ☞ Multiply the numerator and the denominator of $\dfrac{2 + 3i}{1 - 2i}$ by $1 + 2i$ the conjugate of the denominator.
>
> $$\frac{2 + 3i}{1 - 2i} = \frac{2 + 3i}{1 - 2i}\frac{1 + 2i}{1 + 2i} =$$
> $$\frac{2 + 4i + 3i + 6i^2}{1^2 + 2^2} = \frac{-4 + 7i}{5} = \frac{-4}{5} + \frac{7}{5}i.$$

EX 7. Write i^{258} in the form $a + bi$.

> **Solution** ☞ The remainder of the division of 258 by 4 is 2. Therefore, $i^{258} = i^2 = -1$.

EX 8. Write $\sqrt{-3}\,\sqrt{-5}$ in the form $a + bi$.

> **Solution** ☞ $\sqrt{-3}\,\sqrt{-5} = (i\sqrt{3})\,(i\sqrt{5}) = i^2\sqrt{15} = -\sqrt{15}$.
> Be careful $\sqrt{-3} \cdot \sqrt{-5} \neq \sqrt{(-3)(-5)}$.

EX 9. Solve the equation $x^2 + 9 = 0$.

       Solution ☞

$$x^2 + 9 = 0 \Rightarrow x^2 = -9 \Rightarrow$$
$$x = \pm\sqrt{-9} = \pm i\sqrt{9} = \pm 3i.$$

EX 10. Solve the equation $x^2 + 4x + 5 = 0$.

       Solution ☞ Using the quadratic formula, we have

$$x = \frac{-b \pm \sqrt{b^2 - 4 \cdot ac}}{2a} \Rightarrow$$

$$x = \frac{-4 \pm \sqrt{4^2 - 4 \cdot 5}}{2} \Rightarrow$$

$$x = \frac{-4 \pm \sqrt{-4}}{2} \Rightarrow$$

$$x = \frac{-4 \pm 2i}{2} = \frac{2(-2 \pm i)}{2} = -2 \pm i.$$

EX 11. Solve the equation $x^2 - x + 1 = 0$.

       Solution ☞ Using the quadratic formula, we have

$$x = \frac{^-(-1) \pm \sqrt{(-1)^2 - 4 \cdot 1 \cdot 1}}{2 \cdot 1} \Rightarrow$$

$$x = \frac{1 \pm \sqrt{-3}}{2} \Rightarrow$$

$$x = \frac{1 \pm \sqrt{3}\, i}{2}.$$

Chapter 4 Exercises

For problems 1 - 5, rewrite the given expression the form $a + bi$ (where a and b are real numbers).

1. $(-3 - 7i) + (5 + 11i)$

2. $(4 + 3i) - (5 - i)$

3. $(2 - i) \cdot (8 + 9i)$

4. i^{235}

5. $\dfrac{7 - 3i}{1 + 2i}$

Solve equations given on numbers **6 - 9**.

6. $x^2 + 25 = 9$

7. $x^2 + 2x + 6 = 0$

8. $3x^2 - 4x + 5 = 0$

9. $x^2 + 2x + 4 = 0$

For numbers 10 - 15, show all work and print the answer in the space provided.

10. Write $(4i)(5 - 2i)$ in the form $a + bi$.

A.) $-8 - 20i$ B.) $8 - 20i$ C.) $8 + 20i$

D.) $20 - 8i$ E.) None of these.

Answer:_____

11. Write $\dfrac{1}{1 + i} - \dfrac{1}{1 - i}$ in the form $a + bi$.

A.) 0 B.) i C.) $-i$

D.) $\dfrac{1}{2} - \dfrac{1}{2}i$ E.) None of these.

Answer:_____

12. Write $\dfrac{2 + 3i}{4 - 2i}$ in the form $a + bi$.

A.) $\dfrac{1}{2} - \dfrac{3}{2}i$ B.) $\dfrac{7}{20} - \dfrac{4}{5}i$ C.) $\dfrac{3}{4} - \dfrac{5}{8}i$

D.) $\dfrac{1}{10} + \dfrac{4}{5}i$ E.) None of these.

Answer:_____

13. Write $\dfrac{5 - 7i}{-1 + 3i}$ in the form $a + bi$.

A.) $\dfrac{8}{5} + \dfrac{11}{5}i$
B.) $\dfrac{8}{5} - \dfrac{11}{5}i$
C.) $-5 - \dfrac{7}{3}i$
D.) $-\dfrac{13}{5} - \dfrac{4}{5}i$
E.) None of these.

Answer:_____

14. Simplify $\sqrt{-4}\,\sqrt{-8}$

A.) $4i\sqrt{2}$
B.) $-4i\sqrt{2}$
C.) $\sqrt{32}$
D.) $-4\sqrt{2}$
E.) None of these.

Answer:_____

15. Solve the equation $3x^2 - 2x + 5 = 0$ for x.

A.) $x = \dfrac{1 \pm 2i\sqrt{7}}{6}$
B.) $x = \dfrac{1 \pm i\sqrt{14}}{3}$
C.) $x = \dfrac{1 \pm 2i\sqrt{14}}{3}$
D.) $x = \dfrac{1 \pm \sqrt{56}}{2}$
E.) None of these.

Answer:_____

Chapter 5

Quadratic Equations

5.1 Notes about Quadratic Equations

Definition: Any equation that can be written in the form $ax^2 + bx + c = 0$, where a, b, and c are real numbers and $a \neq 0$ is called a *quadratic equation*.

To *solve* a quadratic equation is to find all the values of the unknown that make the equation a true statement.

Every quadratic equation has two solutions. One of the following three statements is true:

   i. if the discriminant $b^2 - 4ac > 0$, the equation has two distinct real solutions.

   ii. if the discriminant $b^2 - 4ac = 0$, the equation has one repeated real solution.

   iii. if the discriminant $b^2 - 4ac < 0$, the equation has two complex solutions.

There are several methods you can use to solve a quadratic equation. These include *Factoring, Completing the Square* and *Quadratic Formula*. The Factoring method works best for quadratic equations that can be factored easily by inspection. The Quadratic Formula or the Completing the Square method will work for all quadratic equations.

5.1.1 Factoring

The idea behind factoring lies in the fact that for two real numbers A and B for which $A \cdot B = 0$, either $A = 0$ or $B = 0$. Therefore, if we can write the original quadratic equation in such a way that it is the product of two linear terms, then the original problem can be simplified to two linear equations. See Examples 1 and 2.

5.1.2   Completing the square

While factoring is a nice method when it is easy to find the factors, not all problems lend themselves to that method. However, all quadratic equations can be solved using the completing the square method or by applying the quadratic formula. See Example 3 for a problem solved by completing the square.

5.1.3   Quadratic formula

The solutions of any quadratic equation of the form $ax^2 + bx + c = 0$, where a, b, and c are real numbers and $a \neq 0$ are given by $x = \dfrac{-b \pm \sqrt{b^2 - 4ac}}{2a}$. See Example 4

5.2   ❖ Errors to Avoid

When using the quadratic formula, be careful to divide both the $-b$-term and the $\sqrt{b^2 - 4ac}$ -term by $2a$. For instance, the solution of $3x^2 + 4x - 2 = 0$ is $\dfrac{-4 \pm \sqrt{4^2 - 4(3)(-2)}}{2(3)}$ **not** $-4 \pm \dfrac{\sqrt{4^2 - 4(3)(-2)}}{2(3)}$.

When solving equations such as $x^2 = 5x$, do not divide both sides by x. Doing so would incorrectly exclude a solution of $x = 0$. For the correct solution, see Example 1.

When solving by factoring, be very careful to avoid the following error:

$x^2 + 8x = 3 \Rightarrow x(x + 8) = 3 \Rightarrow x = 3$ or $x + 8 = 3$. Note that $x = 3$ is NOT a solution of $x^2 + 8x = 3$. The idea $A \cdot B = 0 \Rightarrow A = 0$ or $B = 0$ is only valid when the RHS is 0.

5.3   Examples

EX 1. Solve $x^2 = 5x$ for x.

> **Solution** ☞ To solve this equation, we need to first move all terms with an x in them to the LHS. To do this, we will subtract $5x$ from both sides. This gives $x^2 - 5x = 0$. Factoring an x on the LHS gives $x(x - 5) = 0$. This implies that $x = 0$ and $x - 5 = 0$ or $x = 5$.

EX 2. Solve the equation $x^2 + x - 6 = 0$ for x by Factoring.

> **Solution** ☞ Since $x^2 + x - 6 = 0$ can be written as $(x + 3)(x - 2) = 0$, we can solve this quadratic equation by factoring. Now $(x + 3)(x - 2) = 0$ implies that $x + 3 = 0 \Rightarrow x = -3$ or $x - 2 = 0 \Rightarrow x = 2$. Hence, the solution to the equation $x^2 + x - 6 = 0$ are $x = -3$ and $x = 2$.

> ✓ Checking $x = -3$ in the original equation gives $(-3)^2 + (-3) - 6 = 9 - 3 - 6 = 0$, so $x = -3$ is a solution. Checking $x = 2$ in the original equation gives $(2)^2 + (2) - 6 = 4 + 2 - 6 = 0$, so $x = 2$ is also solution.

EX 3. Solve the equation $4x^2 + 24x - 1 = 0$ by Completing the Square.

> **Solution** ☞ The LHS of this equation cannot be factored with integer coefficients. Hence, it is best to use a method other than Factoring. Here we will solve this equation by Completing the Square.

> First we add 1 to both sides of the equation. We have now $4x^2 + 24x = 1$. Next we divide both sides by 4. This yields $x^2 + 6x = \dfrac{1}{4}$. To complete the square, we add to both sides of the equation the square of half of the x-coefficient. In our case we add $\left(\dfrac{1}{2} \cdot 6\right)^2 = 3^2 = 9$. This gives $x^2 + 6x + 9 = \dfrac{1}{4} + 9$ or $(x + 3)^2 = \dfrac{37}{4}$. Now $x + 3 = \pm\sqrt{\dfrac{37}{4}}$ and $x = -3 \pm \dfrac{\sqrt{37}}{2}$ or $x = \dfrac{-6 \pm \sqrt{37}}{2}$.

EX 4. Use the Quadratic Formula to solve the equation $x^2 - 4x + 1 = 0$.

> **Solution** ☞ The solution to any quadratic equation of the form $ax^2 + bx + c = 0$ is given by the quadratic formula

$$x = \frac{-b \pm \sqrt{b^2 - 4ac}}{2a}$$

> In this example $a = 1, b = -4$ and $c = 1$. Therefore,

$$x = \frac{4 \pm \sqrt{(-4)^2 - 4(1)(1)}}{2(1)} \Rightarrow x = \frac{4 \pm \sqrt{12}}{2} \Rightarrow x = \frac{4 \pm 2\sqrt{3}}{2} \Rightarrow x = \frac{2(2 \pm \sqrt{3})}{2} = 2 \pm \sqrt{3}.$$

EX 5. ✭ (F02#12) For what value(s) of k does the equation $4x^2 + kx + 25 = 0$ have a repeated real solution?

Solution ☞ If the discriminant (the part of the quadratic formula under the radical) is equal to 0, then there is only one repeated real solution to the quadratic equation, and therefore only one distinct solution. The discriminant in general is given by $b^2 - 4ac$ and here $a = 4$, $b = k$ and $c = 25$, so for the discriminant to equal 0, $(k)^2 - 4(4)(25) = 0 \Rightarrow k^2 = 400 \Rightarrow k = \pm 20$.

EX 6. ✭ (S01#19) A parcel of land is 6 ft longer than it is wide and has an area of 216 ft^2. Find the dimensions of the lot.

Solution ☞ Let w be the width of the parcel, its length is $l = W + 6$. The area is $l \cdot w = 216 \Rightarrow (w + 6)w = 216$ (replace l with $w + 6$)

$w^2 + 6w - 216 = 0 \Rightarrow (w - 12)(w + 18) = 0 \Rightarrow$

$w = 12$ or $w = -18$. Since the width cannot be negative, $w = 12$ ft and the length $l = w + 6 = 12 + 6 = 18$ ft.

Chapter 5 Exercises

Solve numbers 1 - 4 by Factoring.

1. $x^2 - 3x - 10 = 0$

$(x + 5)(x - 2) = 0$

$x = -5, 2$

2. $3x^2 - 5x + 5 = 3$ $3x^2 - 5x + 2 = 0$

$(3x - 2)(x - 1) = 0$

$x = 1, 2/3$ $3x - 2 = 0$ $x - 1 = 0$

$3x = 2$ $x = 1$

$x = 2/3$

3. $4x^2 - 9 = 0$

$(2x + 3)(2x - 3)$

$x = -3/2, \ 3/2$ $(\pm 3/2)$

4. $64 - 9t^2 = 0$

$(8 - 3t)(8 + 3t) = 0$

$8 - 3t = 0$ $8 = 3t$ $t = 8/3$

$8 + 3t = 0$ $8 = -3t$ $t = -8/3$

Solve numbers 5 - 9 by Completing the Square.

5. $x^2 - 4x + 2 = 0$

6. $x^2 + 14x = -39$

7. $4x^2 + 4x - 3 = 0$

8. $w^2 = 8w$

9. $x^2 + 4x + 5 = 0$

Use the Quadratic Formula to solve numbers 10 - 14. If there are no real solutions, write "No Real Solution".

10. $3x^2 + 2x - 5 = 0$

11. $2x^2 - 6x = 3$

12. $x^2 + 6x + 9 = 0$

13. $5y = y^2$

14. $2x^2 + 4 = 5x$

For numbers 15 - 19, follow the directions and show all work.

15. For what value(s) of k is $x = 3$ a solution to the equation $x^2 + kx - 15 = 0$.

16. For what value(s) of k is $x = -1$ a solution to the equation $x^2 - 8x + k = 0$.

17. A rectangular parcel of land is 4ft longer than it is wide and has an area of 96ft^2. Find the dimensions of this lot.

18. A rock is thrown straight upward with an initial velocity of 96 feet per second from the top of a 36 feet berm. The number of feet S above the ground after t seconds is given by $S = -16t^2 + 96t + 36$. When will the rock be 100 feet above the ground?

19. Find the radius of a right circular cylinder with no top that has a height of 10cm whose surface area is $125\pi\text{cm}^2$ (note: the surface area of a right circular cylinder is given by $2\pi r h + \pi r^2$).

For numbers 20 - 24, show all work and print the answer in the space provided.

20. Solve the equation $x^2 = 7x$ for x.

   A.) $x = 6$ B.) $x = 6$ and $x = 7$ C.) $x = 0$ and $x = 7$

   D.) $x = 4$ E.) None of these.

Answer:_____

21. Solve the equation $x^2 = 12$ for x.

   A.) $x = \pm 6$ B.) $x = 0$ and $x = 2\sqrt{3}$ C.) $x = 0$ and $x = 6$

   D.) $x = \pm 2\sqrt{3}$ E.) None of these.

Answer:_____

22. Solve the equation $(x + 2)(x - 4) = 6$ for x.

   A.) $x = \dfrac{1 \pm \sqrt{5}}{2}$ B.) $x = 4$ and $x = 10$ C.) $x = -1 \pm \sqrt{15}$

   D.) $x = 1 \pm \sqrt{15}$ E.) None of these.

Answer:_____

43

23. Solve the equation $x^2 - 2x = 6$ for x.

A.) $x = 1 \pm \sqrt{7}$ 　　　　 B.) $x = 2$ and $x = 3$ 　　　　 C.) $x = -1 \pm \sqrt{7}$

D.) $x = 1 \pm \dfrac{\sqrt{7}}{2}$ 　　　　 E.) None of these.

Answer:_____

24. Solve the equation $(2t + 6)^2 - 8 = 0$ for t.

A.) No Real Solution. 　　　　 B.) $t = -5, -1$ 　　　　 C.) $t = -3 \pm \dfrac{\sqrt{2}}{2}$

D.) $t = \dfrac{-6 \pm \sqrt{2}}{2}$ 　　　　 E.) None of these.

Answer:_____

Chapter 6

Other Types of Equations

6.1 Notes about Other Types of Equations

In this chapter, we will extend the techniques for solving linear and quadratic equations to cover other types of equations, including equations of degree greater than 2, radical equations and equations of a quadratic type.

Definition: Equations of higher powers are equations that involve a power greater than 2. For instance, $x^5 + 2x^4 - 4x^3 - 8x^2 = 0$ is an equation of power 5. There is no general formula or method to solve such equations, though it is possible to solve it if we can rewrite it as the product of linear and/or quadratic factors. For the solution, see Example 1.

Definition: *Radical equations* are equations that involve radical expressions. For instance, the problems in Examples 3-5 are examples of radical equations.

To solve radical equation involving only one radical expression (as in Examples 3 and 4), we isolate the radical and then eliminate it by raising both sides of the equation to a power equal to the index of the radical.

In the case where there are two radical expressions in an equation (as in Example 5), we isolate the most complicated radical, then we square both sides. We obtain a simpler equation that can be solved using methods already developed.

6.2 ❖ Errors to Avoid

We need to be careful when solving radical equations. By raising both sides of an equation to a power, extraneous solutions can be introduced. Therefore, though checking solutions into the original equation is always a good idea, it is a **necessary** step in solving radical equations.

6.3   Examples

EX 1. Solve $x^5 + 2x^4 - 4x^3 - 8x^2 = 0$ for x.

Solution ☞ We can factor an x^2-term from each term of the equation obtaining $x^2(x^3 + 2x^2 - 4x - 8) = 0$. Next we can factor $x^3 + 2x^2 - 4x - 8$ by grouping. $x^3 + 2x^2 - 4x - 8 = x^2(x+2) - 4(x+2) = (x+2)(x^2 - 4) = (x+2)(x-2)(x+2)$.

Now $x^2(x+2)(x-2)(x+2) = 0 \Rightarrow$

$$x^2 = 0 \Rightarrow x = 0 \qquad\qquad \text{with multiplicity 2.}$$
$$x + 2 = 0 \Rightarrow x = -2$$
$$x - 2 = 0 \Rightarrow x = 2$$
$$x + 2 = 0 \Rightarrow x = -2$$

So the solutions are $x = 0, -2$, and 2.

EX 2. Solve the equation $2x^5 - 50x^3 = 0$

Solution ☞

$$2x^5 - 50x^3 = 0 \Rightarrow$$
$$2x^3(x^2 - 25) = 0 \Rightarrow$$
$$2x^3(x-5)(x+5) = 0 \Rightarrow$$
$$x = 0, 5 \text{ and } -5.$$

EX 3. Solve $\sqrt[3]{3x - 7} - 2 = 0$ for x.

Solution ☞

$$\sqrt[3]{3x - 7} - 2 = 0 \Rightarrow$$
$$\sqrt[3]{3x - 7} = 2 \Rightarrow$$
$$(\sqrt[3]{3x - 7})^3 = 2^3 \Rightarrow$$
$$3x - 7 = 8 \Rightarrow$$
$$3x = 15 \Rightarrow \qquad x = 5.$$

✐ Checking this solution, $\sqrt[3]{3(5) - 7} - 2 = \sqrt[3]{8} - 2 = 2 - 2 = 0 \Rightarrow$ $x = 5$ is the solution.

EX 4. Solve $\sqrt{2x+5} - 1 = x$ for x.

Solution ☞ To isolate the radical we add 1 to both sides of the equation. $\sqrt{2x+5} = x+1$. Then we raise both sides to the second power obtaining

$$(\sqrt{2x+5})^2 = (x+1)^2 \Rightarrow$$
$$2x + 5 = x^2 + 2x + 1 \Rightarrow$$
$$x^2 - 4 = 0 \Rightarrow$$
$$(x+2)(x-2) = 0 \Rightarrow x = 2 \text{ or } x = -2.$$

✔ To check out solutions, we'll replace x with 2 in the original equation. $\sqrt{2(2)+5} - 1 \Longrightarrow \sqrt{9} - 1 = 2$. Thus, $x = 2$ checks and is a solution of the equation.

Replacing x with -2 in the original equation yields $\sqrt{2(-2)+5} - 1 \Longrightarrow \sqrt{1} - 1 = 0 \neq -2$. Thus, $x = -2$ is NOT a solution and $x = 2$ is the only solution.

EX 5. Solve $\sqrt{2x-1} - \sqrt{x-4} = 2$ for x.

Solution ☞ First add $\sqrt{x-4}$ to both sides of the equation. This yields $\sqrt{2x-1} = 2 + \sqrt{x-4}$. Next we will square both sides of the equation giving

$$\begin{aligned}
(\sqrt{2x-1})^2 &= (2 + \sqrt{x-4})^2 & \Rightarrow \\
2x - 1 &= 4 + 4\sqrt{x-4} + x - 4 \Rightarrow \\
x - 1 &= 4\sqrt{x-4} & \Rightarrow \\
(x-1)^2 &= 16(x-4) & \Rightarrow \\
x^2 - 18x + 65 &= 0 & \Rightarrow \\
(x-5)(x-13) &= 0 & \Rightarrow x = 5 \text{ or } x = 13.
\end{aligned}$$

✔ Checking each of these back into the original equation shows that both $x = 5$ and $x = 13$ are solutions.

EX 6. Solve the equation $x^{2/3} + 7x^{1/3} - 8 = 0$ for x.

Solution ☞ If we let $u = x^{1/3}$, then $u^2 = x^{2/3}$ and we get a quadratic equation in the variable u. This gives

$$u^2 + 7u - 8 = 0 \Rightarrow$$
$$(u-1)(u+8) = 0 \Rightarrow u = 1 \text{ or } u = -8$$

If $u = 1$, then $x^{1/3} = 1$ and $x = 1^3 = 1$.
If $u = -8$, then $x^{1/3} = -8$ and $x = (-8)^3 = -512$.

47

EX 7. Solve the equation $x^4 - 7x^2 + 12 = 0$ for x.

Solution ☞ If we let $u = x^2$, we get the quadratic equation in u $u^2 - 7u + 12 = 0$. We can solve this equation by factoring $u^2 - 7u + 12 = 0 \Rightarrow (u - 3)(u - 4) = 0 \Rightarrow u = 3$ or $u = 4$.
If $u = 3$, then $x^2 = 3$ and $x = \pm\sqrt{3}$.
If $u = 4$, then $x^2 = 4$ and $x = \pm\sqrt{4} = \pm 2$.

EX 8. Solve the equation $\sqrt{x + \sqrt{x + 3}} = 3$ for x.

Solution ☞ To eliminate the 'outside' square root, we square both sides of the equation giving $\left(\sqrt{x + \sqrt{x + 3}}\right)^2 = 3^2 \Rightarrow x + \sqrt{x + 3} = 9$. Isolating the square root on the LHS yields

$$\sqrt{x + 3} = 9 - x \qquad\qquad \Rightarrow$$
$$x + 3 = (9 - x)^2 \qquad\qquad \Rightarrow$$
$$x + 3 = 81 - 18x + x^2 \Rightarrow$$
$$0 = 78 - 19x + x^2 \Rightarrow$$
$$0 = (x - 6)(x - 13) \Rightarrow$$

$x - 6 = 0 \Rightarrow x = 6$ or $x - 13 = 0 \Rightarrow x = 13$.

✎ Checking our potential answers yields:

If $x = 6$, then $\sqrt{6 + \sqrt{6 + 3}} = \sqrt{6 + 3} = 3 \therefore x = 6$ is a solution.

If $x = 13$, then $\sqrt{13 + \sqrt{13 + 3}} = \sqrt{13 + 4} = \sqrt{17} \neq 3 \therefore x = 13$ is not a solution. Therefore, the only solution is $x = 6$.

Chapter 6 Exercises

Solve numbers 1 - 9 for x.

1. $x^3 - 4x = 0$.

2. $x^3 + 2x^2 - 4x - 8 = 0$.

3. $x^4 - 8x^2 + 7 = 0$.

4. $\sqrt[3]{5x - 7} = 0$.

5. $x^4 - 2x^3 + x^2 = 0$.

6. $x + \sqrt{2x + 1} = 7$.

7. $\sqrt{2x + 9} - \sqrt{x + 1} = 2$.

8. $\sqrt[3]{10 - \sqrt{x-1}} = 2$.

9. $x + 3\sqrt{x} - 10 = 0$.

For numbers 10 - 17, show all work and print the answer in the space provided.

10. The solutions to the equation $x^3 - 6x = 0$ are which of the following?

   A.) $x = 0, x = 6$ B.) $x = 3, x = 2, x = -6$ C.) $x = 1, x = \pm 6$

   D.) $x = 0, x = \pm\sqrt{6}$ E.) None of these.

51

Answer:_____

11. The solutions to the equation $t^4 + t^3 - 20t^2 = 0$ are which of the following?

  A.) $t = 0, t = -5, t = 4$ B.) $t = 0, t = -10, t = 10$ C.) $t = 1, t = \pm\sqrt{20}$

  D.) $t = \pm 2, t = \pm\sqrt{5}$ E.) None of these.

Answer:_____

12. Solve the equations $\sqrt{t-1} + 7 = t$ for t.

  A.) $t = 5$ B.) $t = 10$ C.) $t = \dfrac{1 \pm \sqrt{201}}{2}$

  D.) $t = -5, t = 6$ E.) None of these.

Answer:_____

13. Solve the equation $2x + \sqrt{2 - x} = 1$ for x.

    A.) $x = \dfrac{-1 + \sqrt{17}}{x = 8}$ B.) $x = 3/4$ C.) $x = -1/4$

    D.) $x = 1$ E.) None of these.

Answer:_____

14. Solve the equation $\sqrt{2x + 3} - \sqrt{x + 1} = 1$ for x.

    A.) $x = -1, x = 3$ B.) $x = 1$ C.) $x = -3$

    D.) $x = 1/2$ E.) None of these.

Answer:_____

15. Solve the equation $\sqrt[3]{(x-1)^2} = 16$ for x.

    A.) $x = 63, 65$ B.) $x = 63$ C.) $x = 65$

    D.) $x = 8$ E.) None of these.

Answer:_____

16. Solve the equation $x^4 + x^2 - 12 = 0$ for x.

    A.) $x = \pm 2,\ x = \pm i\sqrt{3}$ B.) $x = -4,\ x = 3$ C.) $x = \pm 2,\ \pm 3$

    D.) $x = \pm 2i,\ x = \pm\sqrt{3}$ E.) None of these.

Answer:_____

17. Solve the equation $x^{2/3} - x^{1/3} - 6 = 0$ for x.

    A.) $x = 8$ B.) $x = 27$ C.) $x = 3, x = -2$

    D.) $x = -8, x = 27$ E.) None of these.

Answer:_____

Chapter 7

Linear Inequalities

7.1 Notes about Linear Inequalities

Definition: A *linear inequality* is a mathematical statement involving one variable and at least one of the following four symbols: $\leq$, $\geq$, $>$ or $<$. The *solution of a linear inequality* is the set of all values of the variable that make the statement true. Solutions of inequalities can involve an infinite number of solutions.

Definition: The *graph of a linear inequality* is the set of all points on the real number line that satisfy the inequality.

Some common linear inequalities and their solutions:

The solution of $x > a$ is given by (a, ∞) or

The solution of $x \geq a$ is given by $[a, \infty)$ or

The solution of $a < x \leq b$ is given by $(a, b]$ if $a < b$ (no solution otherwise). The graphical solution if $a < b$ is given.

The solution of $|x| \geq a$ is given by $(-\infty, -a] \cup [a, \infty)$ if $a > 0$ and $(-\infty, \infty)$ otherwise. The solution if $a > 0$ is given.

The solution of $|x| < a$ is given by $(-a, a)$ if $a \geq 0$ and no solution otherwise. The solution if $a > 0$ is given.

7.2 ❖ Errors to Avoid

If when solving an inequality you reach a point at which you will be multiplying or dividing both sides of the inequality by a negative number, this 'switches' the inequality. For instance, $-2x > 4$ is equivalent to $x < -2$.

7.3 Examples

EX 1. ✭ (F03#1) The solution of the inequality $5 - 2x \geq x - 7$ is:

Solution ☞ Collecting all variables one side of $5 - 2x \geq x - 7$ and constants on the other yields $-3x \geq -12$. Dividing each side of the inequality by -3 gives. $x \leq 4$.

In interval notation, this is $(-\infty, 4]$. The graphical solution is below.

$$4$$

EX 2. Solve the inequality $2 < 5x - 3 \leq 6$ and graph its solution on a number line.

Solution ☞ $2 < 5x - 3 \leq 6 \Rightarrow 5 < 5x \leq 9 \Rightarrow$

$1 < x \leq 9/5$.

In interval notation, this is $(-1, 9/5]$. The graphical solution is below.

$$1 \qquad\qquad 9/5$$

EX 3. Solve the inequality $|2x - 5| > 10$ and graph its solution on a number line.

Solution ☞

$$
\begin{array}{ccc}
2x - 5 > 10 & \text{or} & 2x - 5 < -10 \\
2x > 15 & \text{or} & 2x < -5 \\
x > 7.5 & \text{or} & x < -2.5
\end{array}
$$

In interval notation, this is $(-\infty, -2.5) \cup (7.5, \infty)$. The graphical solution is below.

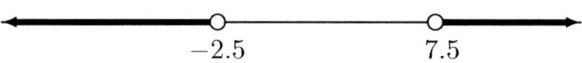
$$-2.5 \qquad\qquad 7.5$$

Name:_____

Section Number:_____

Date:_____

Chapter 7 Exercises

For numbers 1 - 4, graph the given inequality on a number line.

1. $x < -5$.

2. $x \geq 2$.

3. $-2 < x \leq 7$.

4. $|x| < 3$.

For numbers 5 - 16, solve the inequality, if possible, and graph the inequality on a number line.

5. $3x + 4 > 13$.

6. $4 < 6 - 2x$.

7. $7x - 10 \geq 8 - 5x$

8. $5 - 2x \leq 5x + 19$

9. $1 < x + 3 \leq 4$.

10. $2 \leq 6x - 3 < 9$.

11. $5 > 3x - 5 \geq 1$.

12. $7 \geq 8 - 2x > -3$.

13. $|x + 3| \leq 4$.

14. $\left| \dfrac{x + 5}{3} \right| > -2$.

15. $|5 - 2x| < 7$.

16. $\left| \dfrac{x - 4}{2} \right| + 5 \leq 17$.

Show all work for numbers 17 - 18.

17. Car rental company Afids offers a car at a rate of $25 per day, with unlimited mileage. Competing company Bujet offers a car for $15 per day plus $.15 for each mile driven. For what set of miles driven will the Bujet rental car will be the better deal?

18. The Celsius (C) temperature scale is related to the Fahrenheit (F) scale by the equation $F = \dfrac{9}{5}C + 32$. A science experiment is being performed that requires the temperature to be above $55°F$ and below $75°F$. What Celsius temperature range must the experiment be performed within?

For numbers 19 - 21, show all work and print the answer in the space provided.

19. Determine which of the following is a solution of $5x - 3 < 2x + 4$.

    A.) 4   B.) -2 C.) 3   D.) 10   E.) None of these.

    Answer:_____

20. For what values of x is $5 - 7x \geq 6x + 2$?

    A.) $(-\infty, 3]$ B.) $[0, 13/3)$ C.) $[13/3, \infty)$ D.) $[-3, \infty)$ E.) None of these.

    Answer:_____

21. Suppose a and b are real numbers and $ab > 0$. Which of the following must be true about a and b?

    A.) a must be positive.   B.) a and b must have the same sign
    C.) a and b must both be positive.   D.) a and b must have opposite signs
    E.) None of these.

    Answer:_____

Chapter 8

Linear equations in two variables

8.1 Notes about Linear equations in two variables

Definition: An equation of the form $Ax + By + C = 0$, where A, B, and C are numbers and A and B are not both 0, is the *general form* of a linear equation. Any equation that can be put into this form is considered a linear equation.

Definition: The *slope* m between the points (x_1, y_1) and (x_2, y_2) is given by $m = \dfrac{y_2 - y_1}{x_2 - x_1}$.

Definition: An equation of the form $y = mx + b$ is called the *slope-intercept* form of a line. m is the slope of the line and the y-intercept is $(0, b)$.

Definition: An equation of the form $y - y_1 = m(x - x_1)$ is called the *point-slope* form of a line. (x_1, y_1) represents a point on the line, m is the slope of the line.

Note: Each form of an equation of a line uses the variables x and y. These variables represent the different possible values that can be used to make the equation true.

Four important distinctions in lines are the upward sloping, downward sloping, horizontal and vertical. An example of each is plotted below.

Figure 8.1: $y = mx + b$, $m > 0$ Figure 8.2: $y = mx + b$, $m < 0$

Figure 8.3: $y = b$ Figure 8.4: $x = a$

Upward sloping lines have a positive slope and downward sloping lines have a negative slope. *Horizontal lines* have a slope of 0 and therefore have an equation of $y = b$, where b is the y-value of each point on the line. *Vertical lines* are given by the equation $x = a$, where a is the x-value of each point on the line. The slope of a vertical line is undefined.

8.2   Parallel and Perpendicular lines

Two nonvertical lines are *parallel* if and only if their slopes are the same (i.e. $m_1 = m_2$. Any two vertical lines are parallel.

Two lines are *perpendicular* if their slopes are negative reciprocals of each other (i.e. $m_2 = -1/m_1$).

Figure 8.5: Two Parallel Lines   Figure 8.6: Two Perpendicular Lines

8.3 Average Rate of Change

It is often desirable to determine the relative change in a function between two values of the independent variable. This can be done by taking the ratio of the change in the function values to the change in the values of the independent variable. Specifically, the **average rate of change** of a function between $(x_1, f(x_1))$ and $(x_2, f(x_2))$ is given by $\dfrac{f(x_2) - f(x_1)}{x_2 - x_1}$, for $x_2 \neq x_1$. Note that this is equivalent to the slope between two points on the graph or the *slope of the secant line*.

8.4 ❖ Errors to Avoid

When finding the slope between (x_1, y_1) and (x_2, y_2), either of $m = \dfrac{y_2 - y_1}{x_2 - x_1}$ or $m = \dfrac{y_1 - y_2}{x_1 - x_2}$ are acceptable. However, neither $\dfrac{y_2 - y_1}{x_1 - x_2}$ nor $\dfrac{y_1 - y_2}{x_2 - x_1}$ are correct.

8.5 Examples

EX 1. Find the equation of the line through the points $(2, 3)$ and $(7, 2)$ in point-slope form.

> **Solution** ☞ To find the equation of the line through the points, we'll first need to find the slope which is $m = \dfrac{2 - 3}{7 - 2} = \dfrac{-1}{5}$. Since we know the slope and a point, we can now put the line into point-slope form, which gives $y - (3) = \dfrac{-1}{5}(x - 2)$.
>
> Note: either point could be used and $y - (2) = \dfrac{-1}{5}(x - 7)$ is also an acceptable answer.

EX 2. ★ (S01#1) What is the y-intercept of the line passing through the points (-2,6) and (4,-3)?

> A.) 14/3   B.) 5   C.) 3   D.) 9   E.) 6
>
> **Solution** ☞ We first want to find the equation of the line through the points. In particular, if we find the slope-intercept form of the line the y-intercept will be given in the equation. The slope is $m = \dfrac{-3 - 6}{4 - (^-2)} = \dfrac{-9}{6} = \dfrac{-3}{2}$ and the point-slope form of the line is $y - 6 = \dfrac{-3}{2}(x - (^-2))$. Rewriting the expression in slope-intercept form yields $y = \dfrac{-3}{2}x + 3$, so the answer is C.).

EX 3. ✭ (F02#1) Which of the following is the equation of the line through the point $(1, 7)$ and parallel to the line passing through the points $(2, 5)$ and $(-2, 1)$?

A.) $y = x + 7$ B.) $y = 2x + 5$ C.) $y = -2x + 9$

D.) $y = x + 6$ E.) $y = 7x + 1$

Solution ☞ To determine the equation of a parallel line it will be necessary to find the slope of the line through the points $(2, 5)$ and $(-2, 1)$ which is $m = \dfrac{1 - 5}{-2 - 2} = \dfrac{-4}{-4} = 1$. Thus, the equation of the line is $y - 7 = 1(x - 1) \Rightarrow y = x - 6$ or D.).

EX 4. ✭ (S03#17) Find the equation of the line in slope-intercept form that passes through $\left(-\dfrac{1}{2}, 3\right)$ and is perpendicular to the line $y = -\dfrac{2}{5}x + 3$.

Solution ☞ To be perpendicular to $y = -\dfrac{2}{5}x + 3$, the line must have a slope of $x = \dfrac{5}{2}$ and thus the equation of the line must be $y - 3 = \dfrac{5}{2}\left(x - \left(-\dfrac{1}{2}\right)\right)$.

EX 5. Find the average rate of change of $f(x)$ between $x = -1$ and $x = 3$ using the graph of $y = f(x)$ below.

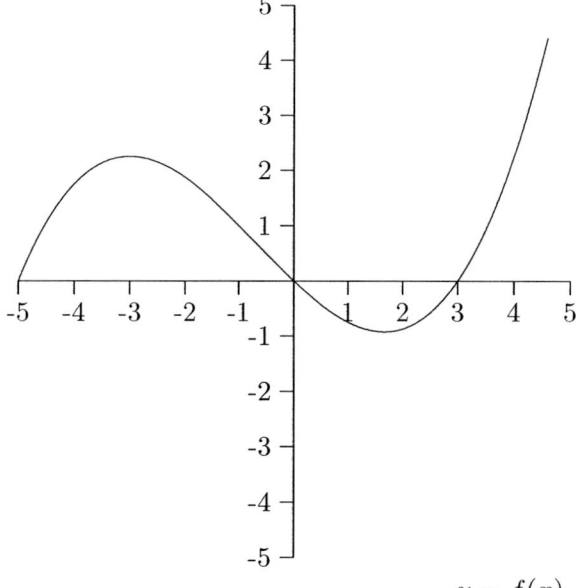

$$y = f(x)$$

Solution ☞ Here we see that $f(-1) = 1$ and $f(3) = 0$, so the average rate of change is $\dfrac{0 - 1}{3 - 1} = -\dfrac{1}{2}$.

Chapter 8 Exercises

Find the equation of the line in point-slope form through the given points for problems 1 - 4.

1. $(1, 2)$ and $(6, 12)$.

2. $(10, 2)$ and $(8, 4)$.

3. $(-1, 3)$ and $(2, -7)$.

4. $(1/2, 4/5)$ and $(6, 9)$.

Find the equation of the line in slope-intercept form through the given points for problems 5 - 9.

    5. $(2, 5)$ and $(6, 1)$.

    6. $(-1, 2)$ and $(3, 18)$.

    7. $(-4, 2)$ and $(5, 12)$.

    8. $(3, 1/2)$ and $(8, -7)$.

9. $(1, 4)$ and $(12, 4)$.

10. Determine the value of the x-coordinate of the point $(h, 5)$ that lies on the line $y = 3x - 7$.

11. Find the equation of the line parallel to $y = -2x + 7$ that passes through the point $(5, 10)$.

12. Find the equation of the line perpendicular to $5x - 10y = 20$ that passes through the point $(1, -1)$.

13. Determine the equation of the lines parallel and perpendicular to $y = 4x + 7$ that pass through the point $(5, -2)$.

14. The value of a certain type of car loses value linearly over time. The original sales price is $22,000 and the price after three years is $14,500 and the car continues to lose value at the same rate for four more years. How long after the car was purchased will the car be worth $5,000?

15. Find the average rate of change of the function $f(x) = -x^3 + 5x$ from $x = -2$ to $x = 1$.

16. Find the average rate of change of the function $g(x) = (x - 2)^2 + 5$ from $x = -1$ to $x = 5$.

For numbers 17 - 21, show all work and print the answer in the space provided.

17. The slope of the line through the points $(3, -2)$ and $(5, 10)$ is

    A.) 4   B.) 1/6   C.) 1

    D.) 6   E.) None of these.

Answer:_____

18. The equation of the line through the points $(1, -2)$ and $(6, 8)$ is which of the following?

    A.) $x + 2y = -3$ B.) $y = 2x + 4$ C.) $2x - y = 4$

    D.) $y = \left(\dfrac{1}{2}\right)x + 5$ E.) None of these.

Answer:_____

19. The equation of the line parallel to $x = -3$ and passing through the point $(-1, 4)$ is which of the following?

A.) $y = \left(\frac{1}{3}\right)x + \frac{13}{3}$ B.) $x = -1$ C.) $y = 4$

D.) $x = 1$ E.) $x = 4$

Answer:_____

20. The x-intercept of the line passing through the points $(1, 2)$ and $(6, -5)$ is which of the following?

A.) 1   B.) $-3/7$ C.) 7/3   D.) 13/5   E.) 17/7

Answer:_____

21. The equation of the line perpendicular to $2x - 5y = 5$ that passes through the point $(-1, 3)$ is

A.) $2x - 5y = -1$ B.) $5x + 2y = 1$ C.) $5x + 2y = -1$

D.) $5x - 2y = 1$ E.) None of these.

Answer:_____

Chapter 9

Functions

9.1 Notes about Functions

Definition: A *function* f is a rule that assigns to each element x in a set A exactly one element $f(x)$ in another set B. $f(x)$ represents the value of the function f for the given value x and $f(x)$ is read "f of x." $f(x)$ is also called the image of x under f. The set A is called the *domain* and the set B is called the codomain. The *range* of f is the set of all values of the function f within B that can occur using values from the set A.

Definition: If y is a function of x, x is called the *independent variable* and y is called the *dependent variable*.

Note that while it is common to denote functions with one letter (i.e. f, g or h), this need not be the case. The common logarithm is a function that is written $\log(x)$. Some common trigonometric functions are written $\sin(x)$, $\cos(x)$ and $\tan(x)$.

Definition: A *piecewise-defined function* is a function made of several different parts, each with a specified domain.

Definition: Some functions are given with a stated domain, such as $f(x) = x + 2$, for $x > 5$. Here the domain is $(5, \infty)$. Others have what is called an *implied domain*. For instance, $g(x) = \sqrt{x}$ has the implied domain over the codomain of the real numbers of $x \geq 0$ or $[0, \infty)$. When asked to find the domain, it is typically the implied domain that you are asked to find. Unless told otherwise, assume the codomain to be the real numbers.

Definition: Recall that the slope between two points (x_1, y_1) and (x_2, y_2) is given by $m = \dfrac{y_2 - y_1}{x_2 - x_1}$. If the two points are on the graph of a function, it is often useful to consider the slope formula in terms of that function. In particular, if we know that $y = f(x)$ and we call x_1 a and we consider the distance between x_1 and x_2 to be h, then the points can be expressed as $(a, f(a))$ and $(a+h, f(a+h))$. Substituting into the slope formula yields $m = \dfrac{y_2 - y_1}{x_2 - x_1} = \dfrac{f(a+h) - f(a)}{(a+h) - (a)} = \dfrac{f(a+h) - f(a)}{h}$, for $h \neq 0$. This last formula is called the *difference quotient*.

9.2 ❖ Errors to Avoid

Be careful that you don't confuse the notation $f(x)$ to mean that f is being multiplied times x. Again, f is a *rule* that acts on x.

9.3 Examples

EX 1. Explain in words what the function $g(x) = 3x - 1$ means.

> **Solution** ☞ $g(x) = 3x - 1$ means that each value that x takes will be multiplied by three and then decreased by one. The implied domain of the function is all real numbers, often denoted $\mathbb{R}$, because any real number can be multiplied by 3 and then have 1 subtracted from it.

EX 2. Evaluate $f(x) = \sqrt{4 - x}$ at $x = -5, 0, 4$, and 5.

Solution ☞

$$f(-5) = \sqrt{4 - ^- 5} = \sqrt{9} = 3$$
$$f(0) = \sqrt{4 - 0} = \sqrt{4} = 2$$
$$f(4) = \sqrt{4 - 4} = \sqrt{0} = 0$$
$$f(5) = \sqrt{4 - 5} = \sqrt{-1} = \text{undefined (over the real numbers)}$$

EX 3. ★ (S02#19) For the piecewise function

$$f(x) = \begin{cases} 1 & \text{if } x < 0, \\ x & \text{if } 0 \leq x \leq 2, \\ x^2 - 2 & \text{if } x > 2. \end{cases}$$

(a) $f(-1) =$

Solution ☞ To determine this, first determine which portion of the piecewise defined function the x-value is located in. In this instance, -1 is in the $x < 0$ category and thus should be evaluated using the rule $f(x) = 1$. Thus, $f(-1) = 1$.

(b) $f(0) =$

Solution ☞ Because 0 is located in the $0 \leq x \leq 2$ category, we will use the rule $f(x) = x$. Thus, $f(0) = 0$.

(c) $f(4) =$

Solution ☞ Because 4 is located in the $x > 2$ category, we will use the rule $f(x) = x^2 - 2$. Thus, $f(4) = (4)^2 - 2 = 14$.

EX 4. Find the domain of $g(x) = \dfrac{\sqrt{x+9}}{2x-4}$.

Solution ☞ The domain is the set of all x-values for which the function is defined. Over the real numbers, this will be the set of values such that the value under the square root is non-negative and the denominator is non-zero. To find this, we will set $x + 9 \geq 0$ and recognize that $2x - 4 \neq 0$. Solving each expression gives $x \geq -9$ and $x \neq 2$.

To help correctly combine these two expressions, let's look at a graph of each expression separately.

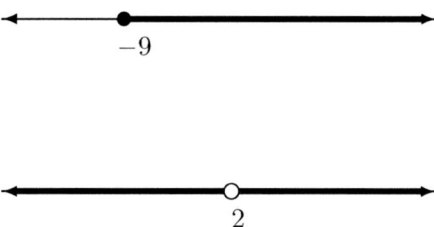

Combining together these expressions into one requires that we use only the portion of the real line that is shaded giving the following:

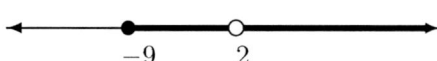

From the graph of the inequality, we can see that the domain is $[-9, 2) \cup (2, \infty)$.

EX 5. Use $f(x) = 3x^2 - 8$ to answer the questions that follow.

(a) Find the slope between $(1, f(1))$ and $(5, f(5))$.

Solution ☞ $f(1) = 3(1)^2 - 8 = -5$ and $f(5) = 3(5)^2 - 8 = 67$, so the slope between the two points is $m = \dfrac{67 - {}^-5}{5 - 1} = \dfrac{72}{4} = 18$.

(b) Find the slope between $(1, f(1))$ and $(3, f(3))$.

Solution ☞ $f(1) = 3(1)^2 - 8 = -5$ and $f(3) = 3(3)^2 - 8 = 19$, so the slope between the two points is $m = \dfrac{19 - {}^-5}{3 - 1} = \dfrac{24}{2} = 12$.

(c) Find the slope between $(1, f(1))$ and $(2, f(2))$.

Solution ☞ $f(1) = 3(1)^2 - 8 = -5$ and $f(2) = 3(2)^2 - 8 = 4$, so the slope between the two points is $m = \dfrac{4 - {}^-5}{2 - 1} = \dfrac{9}{1} = 9$.

Alternate Solution ☞ Using the difference quotient can make this problem easier if you notice that each problem is asking for the slope between $(1, f(1))$ and another point. Choosing 1 as our a gives the difference quotient as: $\dfrac{f(1 + h) - f(1)}{h} = \dfrac{(3(1 + h)^2 - 8) - (3(1)^2 - 8)}{h} = \dfrac{(3 + 6h + 3h^2 - 8) - {}^-5}{h} = \dfrac{6h + 3h^2}{h} = 6 + 3h$ for $h \neq 0$. Thus, for (a) where $h = 4$ (because $5 - 1$ gives 4) the slope is $6 + 3(4) = 18$. For (b), where $h = 2$, the slope is $6 + 3(2) = 12$. For (c), where $h = 1$, the slope is given by $6 + 3(1)$. Notice that each of these values agrees with what was found previously.

EX 6. ★ (S05#20) If $f(x) = 3x^2 - 5$.

(a) Evaluate $f(2)$.

Solution ☞ $f(2) = 3(2)^2 - 5 = 3(4) - 5 = 12 - 5 = 7$.

(b) Simplify $f(a + 1)$.

Solution ☞ $f(a + 1) = 3(a + 1)^2 - 5 = 3(a^2 + 2a + 1) - 5 = 3a^2 + 6a + 3 - 5 = 3a^2 + 6a - 2$.

(c) Simplify $\dfrac{f(x + h) - f(x)}{h}$.

Solution ☞ $\dfrac{3(x + h)^2 - 5 - (3x^2 - 5)}{h} = \dfrac{3(x^2 + 2xh + h^2) - 5 - 3x^2 + 5}{h}$

$\dfrac{3x^2 + 6xh + 3h^2 - 5 - 3x^2 + 5}{h} = \dfrac{6xh + 3h^2}{h}$

$\dfrac{h(6x + 3h)}{h} = 6x + 3h$ or $3(2x + h)$.

Chapter 9 Exercises

For numbers 1 - 6, evaluate the given functions at the given values, if possible.

  1. Evaluate $f(x) = 2x - 7$ when $x = -3, 0, 1$, and 6.

  2. Evaluate $g(x) = 3x - x^2$ when $x = -4, -1, 0, 2$, and 5.

  3. Evaluate $h(x) = \dfrac{\sqrt{4 - x^2}}{x - 1}$ when $x = -7, -2, 0, 1$, and 4.

  4. Evaluate $r(t) = \dfrac{3t}{5 - t}$ when $t = -5, 0, 2$, and 5.

5. Evaluate $f(x)$ (below) when $x = -3, -1, 0, 4$, and 7.

$$f(x) = \begin{cases} 2x - 4 & \text{if } x < -2 \\ 3 & \text{if } -2 \le x \le -1 \\ 7 - x & \text{if } -1 < x \le 2 \\ x^2 & \text{if } x > 2 \end{cases}$$

6. Evaluate $g(x)$ (below) when $x = -5, -3, 0, 1, 2$, and 6.

$$g(x) = \begin{cases} 0 & \text{if } x < -3 \\ 4 - x & \text{if } -2 \le x \le -1 \\ 5 & \text{if } -1 < x < 0 \\ 3 + x^2 & \text{if } 0 \le x < 1 \\ \sqrt{x + 3} & \text{if } x \ge 2 \end{cases}$$

For numbers 7 - 13, find the domain of the given functions.

7. $f(x) = 3x + 19$.

8. $g(x) = 5x^4 - 2x^3 + 7x - 495$.

9. $f(x) = \sqrt{x+1}$.

10. $g(x) = \dfrac{1}{x}$.

11. $h(x) = \sqrt{19 - x}$.

12. $f(s) = \dfrac{2s}{\sqrt{s+1}}$.

13. $g(t) = \dfrac{\sqrt{2t+7}}{t^3 - 1}$.

14. Find and simplify the difference quotient for $g(t) = t^3 - 7$ around $t = a$.

15. Use $f(x) = 7 - 3x + x^2$ to answer the questions that follow.

   (a) Find and simplify the difference quotient for $f(x)$ using $a = 3$.

   (b) Use your answer to (15a) to find the slope between the points $(3, f(3))$ and $(6, f(6))$.

   (c) Use your answer to (15a) to find the slope between the points $(3, f(3))$ and $(4, f(4))$.

   (d) Use your answer to (15a) to find the slope between the points $(3, f(3))$ and $(0, f(0))$.

For numbers 16 - 18, show all work and print the answer in the space provided.

16. For which of the functions below does $g(a) + g(h) = g(a + h)$?

   A.) $g(x) = x^2$ B.) $g(x) = 1 + x$ C.) $g(x) = x^3$

   D.) $g(x) = 7x$ E.) All of these.

Answer:_____

17. Which of the following represents a function from the set R to the set T?

A.)

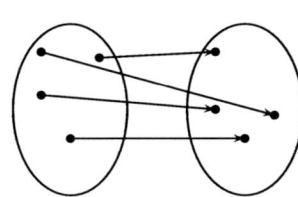

R T

B.)

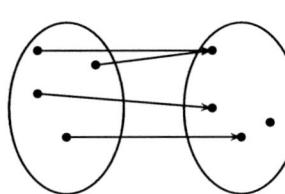

R T

C.)

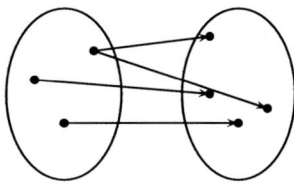

R T

D.)

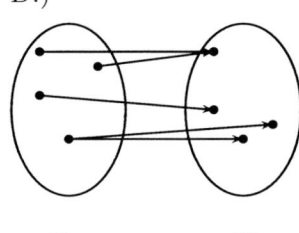

R T

E.)

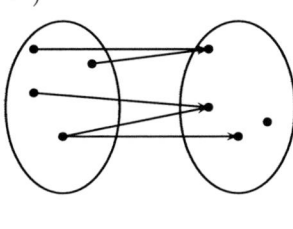

R T

Answer:_____

18. Which of the following represents the domain D and range R of $y = f(x) = 3x - 2$, $-3 \leq x < 6$.

A.) $D = -3 \leq x < 6, R = -3 \leq y < 6$

B.) $D = \mathbb{R}, R = \mathbb{R}$

C.) $D = -1/3 \leq x < 8/3, R = -3 \leq y < 6$

D.) $D = -3 \leq x < 6, R = -11 \leq y < 16$

E.) None of these.

Answer:_____

Chapter 10

Graphs of Functions

10.1 Notes about Graphs of Functions

Definition: The graph of a function f is the collection of points $(x, f(x))$ for all values of x in the domain of f.

Recall that a function f is a rule that assigns to each element x in a set A exactly one element $f(x)$ in another set B. Thus, if for one x-value, there exists more than one $f(x)$ or y-value, then $y = f(x)$ is not a function. This idea is represented by the vertical line test.

Vertical Line Test: A graph representing a function can have no more than one $f(x)$-value for each x-value. This means that a graph represents a function if and only if no vertical line intersects the graph in more than one point.

Definition: A function is *increasing* on an interval I if $f(x_2) > f(x_1)$ whenever $x_2 > x_1$ for each x-value x_1 and x_2 in the interval. Graphically, a function is increasing on I if, while viewed from left to right, it is going up.

Definition: A function is *decreasing* on an interval if $f(x_2) < f(x_1)$ whenever $x_2 > x_1$ for each x-value x_1 and x_2 in the interval. Graphically, a function is decreasing on I if, while viewed from left to right, it is going down.

Definition: A point $(a, f(a))$ on the graph of $y = f(x)$ is called the *relative maximum* if $f(a) \geq f(x)$ for all x such that $c < a < b$ and $c < x < b$, for some values c and b in the domain of f.

Definition: A point $(a, f(a))$ on the graph of $y = f(x)$ is called the *relative minimum* if $f(a) \leq f(x)$ for all x such that $c < a < b$ and $c < x < b$, for some values c and b in the domain of f.

Definition: Collectively, the maxima and minima are called the extrema.

Definition: The *zeros* (or *roots*) of a function $f(x)$ are the x-values for which $f(x) = 0$. If a is a zero of f, then $(a, 0)$ is an x-intercept of the graph of $y = f(x)$.

Definition: The graph of a relation is *symmetric* about the:

- x-axis if the point $(a, -b)$ is on the graph when (a, b) is on the graph. An equation with x-axis symmetry is unchanged when $-y$ replaces y.

- y-axis if the point $(-a, b)$ is on the graph when (a, b) is on the graph. An equation with y-axis symmetry is unchanged when $-x$ replaces x.

- *origin* if the point $(-a, -b)$ is on the graph when (a, b) is on the graph. An equation with origin symmetry is unchanged when $-x$ and $-y$ replace x and y, respectively.

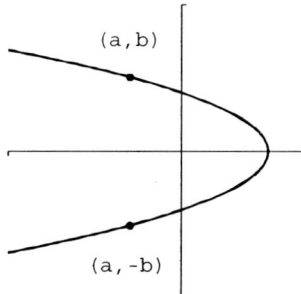

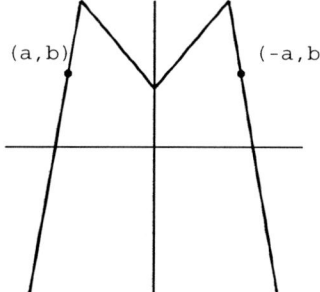

Figure 10.1: Symmetric about x-axis.   Figure 10.2: Symmetric about y-axis.

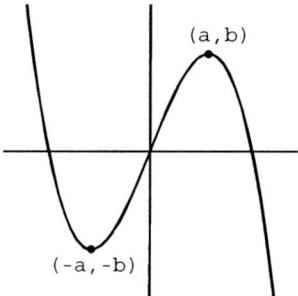

Figure 10.3: Symmetric about origin.

Definition: A function f is *even* if $f(x) = f(-x)$ for all x in the domain of f. The graph of an even function is symmetric about the y-axis.

Definition: A function f is *odd* if $f(x) = -f(-x)$ (or equivalently $f(-x) = -f(x)$ for all x in the domain of f). The graph of an odd functions is symmetric about the origin.

10.2 ❖ Errors to Avoid

It is possible for a function to have no relative maximum or relative minimum over an interval. It is also possible for a function to be neither even nor odd; many are neither (see Examples 5.*i* - .*iii*).

The intervals a function is increasing or decreasing over are described using x-values. For instance, if a function is said to be decreasing from $(-1, 2)$ means that the graph of the function is going for the x-values between -1 and 2. These numbers do not represent a point or a set of y-values.

10.3 Examples

EX 1. ★ (F01#14) Consider the graphs below. Which of the curves is not the graph of a function?

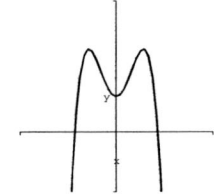

i.

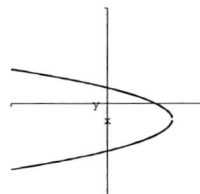

ii.

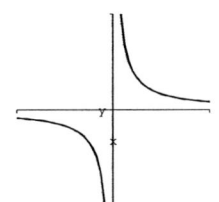

iii.

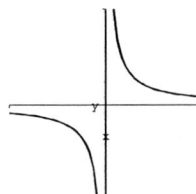

iv.

Solution ☞ With regard to the vertical line test, each of *i*), *iii*) and *iv*) pass and *ii*) does not. Thus, *ii*) is not a function.

EX 2. Find the zeros of the following functions:

(a) $f(x) = 5x - 10.$

Solution ☞ To find the zeros, we must determine where the function is equal to 0. Setting $5x - 10 = 0 \Rightarrow 5x = 10 \Rightarrow x = 2.$

(b) $g(x) = 3x^2 - 7x + 4.$

Solution ☞ Setting $g(x) = 0$ gives $0 = 3x^2 - 7x + 4 \Rightarrow 0 = (3x - 4)(x - 1) \Rightarrow 3x - 4$ or $x - 1 = 0 \Rightarrow x = 4/3$ or $x = 1.$

EX 3. Find the values asked for below based on the graph of $y = f(x)$ in Figure 10.4.

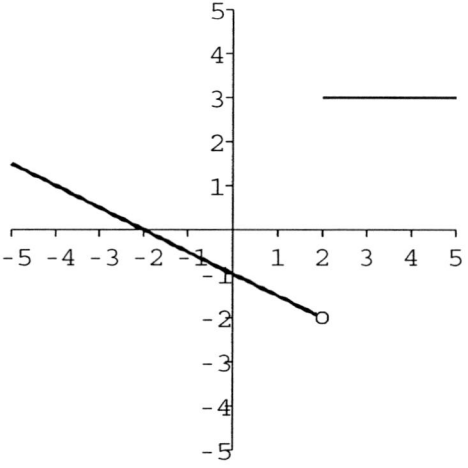

Figure 10.4: $y = f(x)$

(a) $f(-1) =$

Solution ☞ Looking at the graph, we see that when $x = -1$, $y = -1/2$, so $f(-1) = -1/2.$

(b) $f(2) =$

Solution ☞ Looking at the graph, we see that when $x = 2$, $y = 3$, so $f(2) = 3.$

(c) $f(4) =$

Solution ☞ Looking at the graph, we see that when $x = 4$, $y = 3$, so $f(4) = 3.$

EX 4. Determine where the graph of the function below is increasing and where it is decreasing and find the relative maxima and minima.

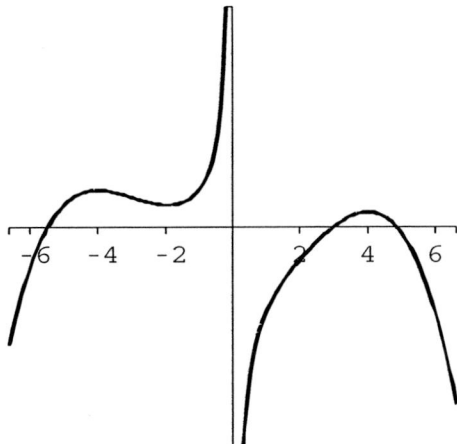

Figure 10.5: $y = f(x)$

Solution ☞ Looking at the graph of $y = f(x)$, starting on the left and working to the right, we notice that the graph is increasing until it gets to $x = -4$, it is decreasing between $x = -4$ and $x = -2$, then increases from $x = -2$ until it gets to $x = 0$, where it is undefined. From $x = 0$ until $x = 4$ the function increases and then decreases from $x = 4$ on.

So the function is increasing on $(-\infty, -4) \cup (-2, 0) \cup (0, 4)$ and is decreasing on $(-4, -2) \cup (4, \infty)$. Note that the we did not say that the graph is increasing on $(-2, 4)$ because the function is undefined when $x = 0$.

Noting where the graph is increasing, we see that the three x-values that could be the relative maxima are at $-4, 0$, and 4. Looking at the graphs, we see that the relative maxima are at $(-4, f(-4))$ and $(4, f(4))$.

Noting where the graph is decreasing, we see that there is only one x-value that could be a relative minimum is $x = -2$. From the graph, we see that the point $(-2, f(-2))$ is a relative minimum.

EX 5. For the following graphs,

(a) determine the domain and range,

(b) find the relative maxima and minima,

(c) find the zeros, and

(d) determine if the graph is even, odd or neither.

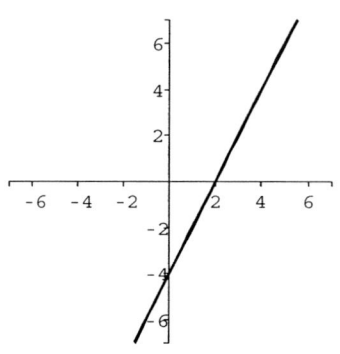

i.

Solution ☞

(a) The domain of this graph is all real numbers because it continues to both the left and the right. The range seems to be all real numbers because the graph continues both upward and downward.

(b) Because the graph is always increasing, it has neither a relative maximum nor a relative minimum.

(c) The graph crosses the x-axis when $x = 2$, so it has a zero at $x = 2$.

(d) Because the graph does not display y-axis or origin symmetry, the function is neither even nor odd.

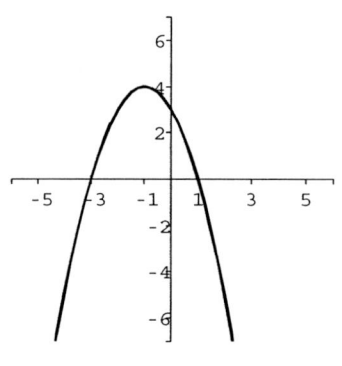

ii.

Solution ☞

(a) The domain of this graph is all real numbers because it continues both to the left and the right. The graph goes no higher vertically than 4 but continues downward, so the range is $(-\infty, 4]$.

(b) The graph achieves a relative (and absolute) maximum of 4 when $x = -1$. It has no relative minimum.

(c) The graph has zeros at $x = -3$ and $x = 1$.

(d) Because the graph does not display y-axis or origin symmetry, the function is neither even nor odd.

Solution ☞

(a) The domain is $[-6, 5]$. To see this it is helpful to scan from left to right, noticing that each value between -6 and 5 is taken, as well as -6 and 5. Scanning from the bottom of the graph upwards yields a range of $[-8, -2) \cup [0, 8]$.

iii.

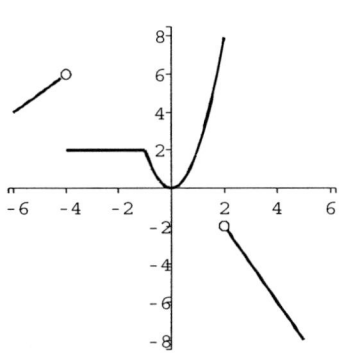

(b) $f(x) = 4, 0$, and -8 are relative minima which occur when $x = -6, 0$, and 5. $f(x) = 8$ is a relative maxima that occurs when $x = 2$. Note $f(x) = 2$ is both a relative maximum AND a relative minimum over the interval $[-4, -1]$. Also, there is no relative maximum at $(-4, 6)$ or $(-2, 2)$ because the function does not include these points.

(c) The function has a zero only at $(0, 0)$.

(d) The function is neither even nor odd.

Solution ☞

(a) The domain is $\mathbb{R}$ and the range is $\mathbb{R}$.

(b) A relative minimum occurs at $(1, -3)$ and a relative maximum occurs at $(-1, 3)$.

(c) The function has zeros when $x = -2, 0$, and 2.

(d) Because the graph displays origin symmetry, the function is odd.

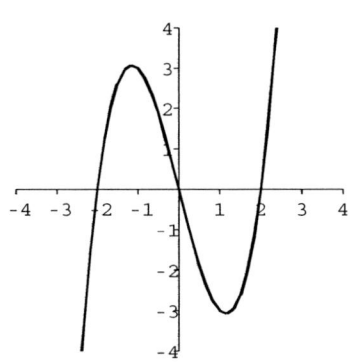

iv.

Chapter 10 Exercises

For numbers 1 - 4, find the zeros of the functions algebraically.

1. $f(x) = 4x - 3$

2. $g(x) = 10 - 3x + x^2$

3. $h(x) = \dfrac{3 - 5x^2}{x^3 + 4x^2 - 21x}$

4. $q(t) = t^4 - 3t^2 + 2$

Use the graphs at right to answer questions 5 and 6.

5. Find the following values of f.

    (a) $f(-4)=$_____

    (b) $f(-1)=$_____

    (c) $f(0)=$_____

    (d) $f(2)=$_____

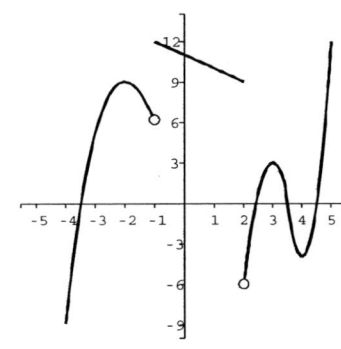

Graph of $y = f(x)$.

6. Find the following values of $g(x)$.

(a) $g(-5)=$_____

(b) $g(0)=$_____

(c) $g(1)=$_____

(d) $g(4)=$_____

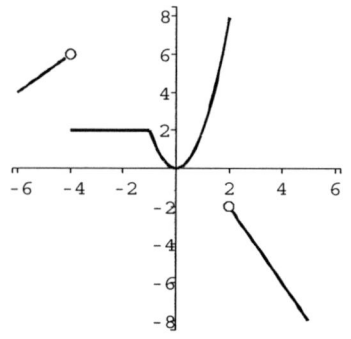

Graph of $y = g(x)$.

Answer questions 7 - 9 on the graphs provided.

7. Sketch the graph of a function that is odd and increasing on $(0, \infty)$.

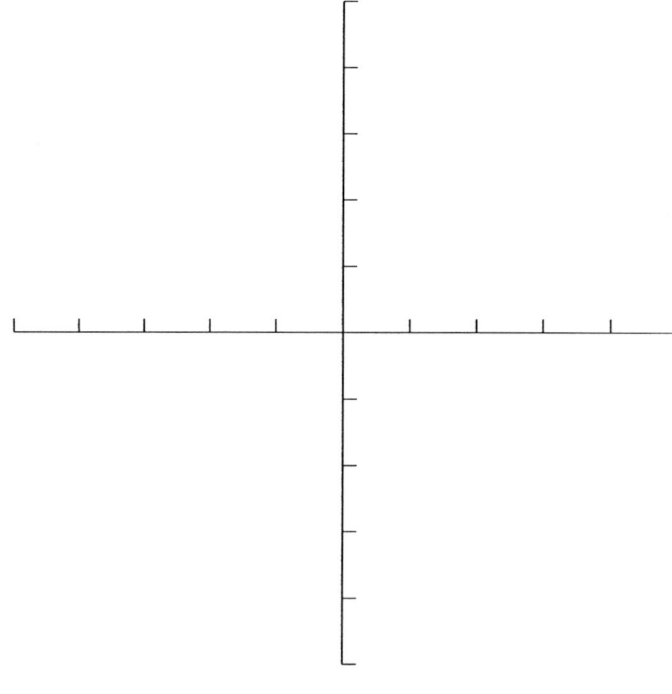

8. Sketch the graph of a function that is even, decreasing on $(-\infty, -4)$, increasing on $(-4, 0)$ and has roots at $x = -5$ and -1.

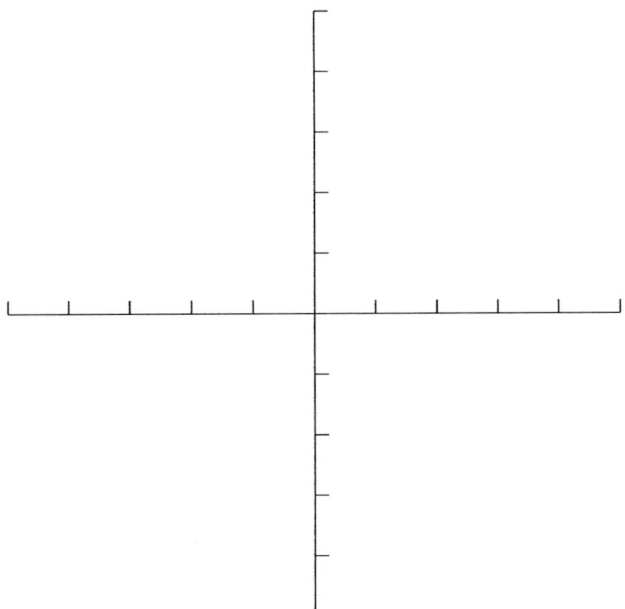

9. Sketch the graph of a function that has relative maxima at $x = -2$ and 3 only and relative minima at $x = -4$ only and zeros at $x = -3, 0$ and 5 only.(Hint: not all functions are continuous.)

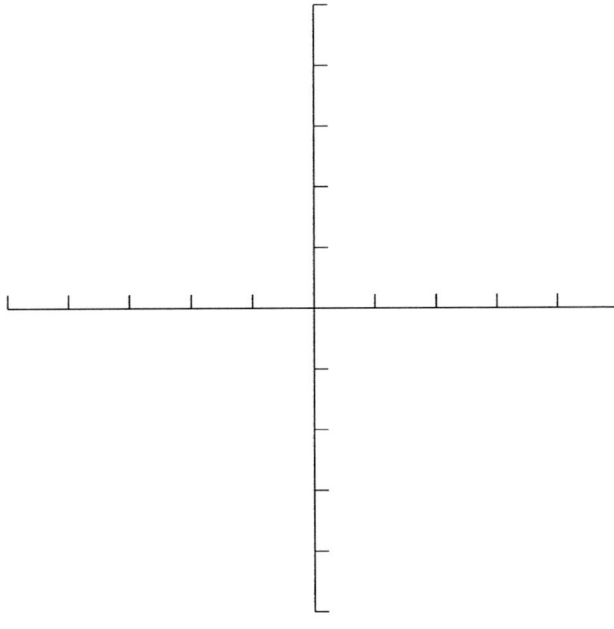

Use Figure 10.6 to answer questions 10 - 12.

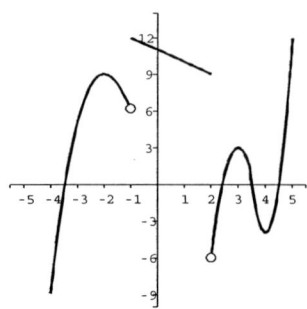

Figure 10.6: Graph of $y = f(x)$

10. Determine the domain D and range R of the function $f(x)$.

    A.) $D : [-5, 5]; R : [-12, 12]$ B.) $D : [-4, 5]; R : [-9, 12]$

    C.) $D : (-1, 2); R : (-6, 6)$ D.) $D : [-4, -1) \cup (-1, 2) \cup (2, 5]; R : [-9, 12]$

    E.) None of these.

    Answer:_____

11. Find the relative minima of the function $f(x)$.

    A.) $(4, -4)$ B.) $(-1, 6), (2, -6), (4, -4)$

    C.) $(-4, -9), (4, -4)$ D.) $(-2, 9), (-1, 12), (3, 3), (5, 12)$

    E.) $(-4, -9), (4, -4)$

    Answer:_____

12. Determine the interval over which $f(x)$ is decreasing.

    A.) $(-2, -1) \cup (-1, 2) \cup (3, 4)$ B.) $(9, 6) \cup (12, 9) \cup (3, -4)$

    C.) $(-2, -1) \cup (-1, 4)$ D.) $(-2, -1) \cup (3, 4)$

    E.) $(-4, -2) \cup (2, 3) \cup (4, 5)$

    Answer:_____

13. Determine which of the following is the graph of an odd function.

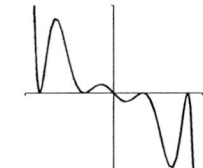

A.

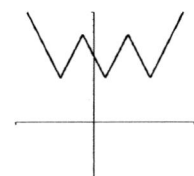

B.

C.

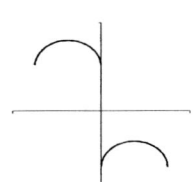

D.

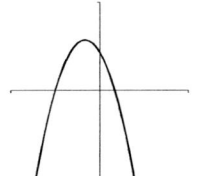

E.

Answer:_____

14. Determine which of the following is the graph of an even function.

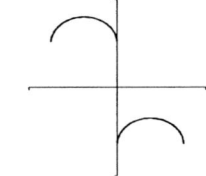

A.

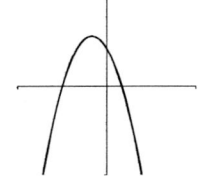

B.

C.

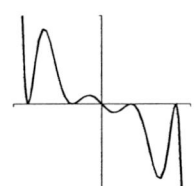

D.

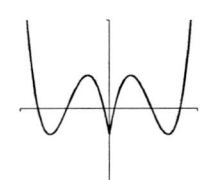

E.

Answer:_____

15. Determine which of the following is the graph of a function.

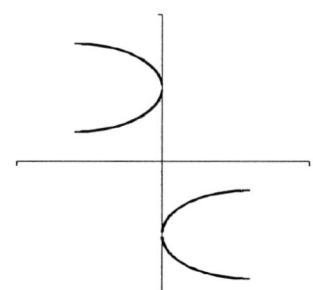

A.)

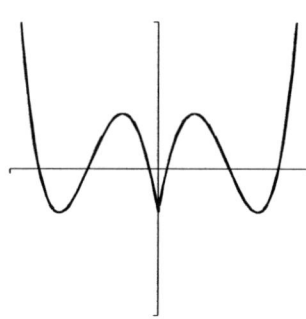

B.)

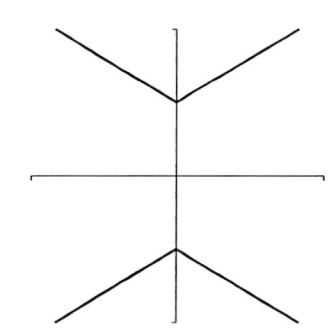

C.)

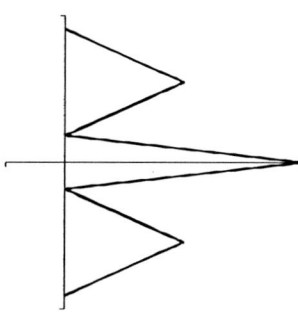

D.)

E.) None of these.

Answer:_____

For problems 16 - 19, determine whether the indicated relation has x-axis, y-axis, origin, or none of these symmetries. Show all work and print the answer in the space provided.

16. $y = 5x^3 - 2x$.

   A.) x-axis symmetry only   B.) y-axis symmetry only
   C.) origin symmetry only   D.) x-axis and y-axis symmetry only
   E.) None of these.

Answer:_____

17. $y = x^4 + 9$.

    A.) x-axis symmetry only   B.) y-axis symmetry only

    C.) origin symmetry only   D.) x-axis and y-axis symmetry only

    E.) None of these.

Answer:_____

18. $y = 7x + 3$.

    A.) x-axis symmetry only   B.) y-axis symmetry only

    C.) origin symmetry only   D.) x-axis and y-axis symmetry only

    E.) None of these.

Answer:_____

19. $y^2 = 2x - 6$.

    A.) x-axis symmetry only   B.) y-axis symmetry only

    C.) origin symmetry only   D.) x-axis and y-axis symmetry only

    E.) None of these.

Answer:_____

Chapter 11

Some important functions

11.1 Notes about Some important functions

The graphs below will be used in many different contexts. You will be expected to know these basic graphs.

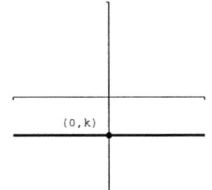

Figure 11.1: Constant Function $f(x) = k$

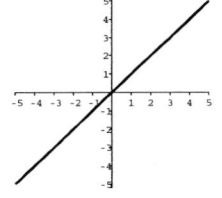

Figure 11.2: Identity Function $f(x) = x$

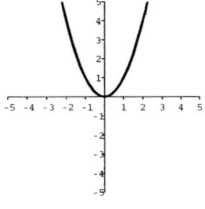

Figure 11.3: Quadratic Function $f(x) = x^2$

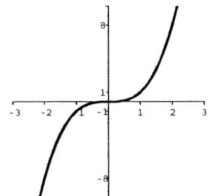

Figure 11.4: Cubic Function $f(x) = x^3$

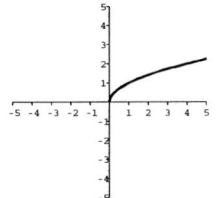

Figure 11.5: Square Root Function $f(x) = \sqrt{x}$

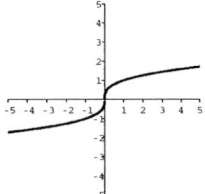

Figure 11.6: Cube Root Function $f(x) = \sqrt[3]{x}$

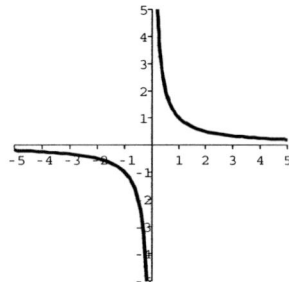

Figure 11.7:   Reciprocal Function
$f(x) = \dfrac{1}{x}$

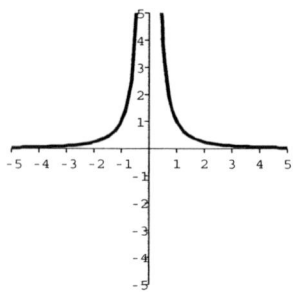

Figure 11.8: Square of the Reciprocal
Function $f(x) = \dfrac{1}{x^2}$

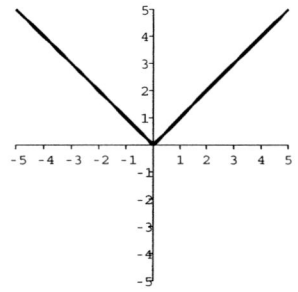

Figure 11.9: Absolute Value Func-
tion $f(x) = |x|$

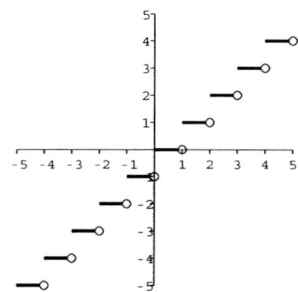

Figure 11.10: Greatest Integer Function $f(x) = \lfloor x \rfloor$

The piecewise definitions of both $|x|$ and $\lfloor x \rfloor$ are below.

$$|x| = \begin{cases} -x, & x < 0 \\ x, & x \geq 0 \end{cases} \qquad \lfloor x \rfloor = \begin{cases} \vdots & \\ -2, & -2 \leq x < -1 \\ -1, & -1 \leq x < 0 \\ 0, & 0 \leq x < 1 \\ 1, & 1 \leq x < 2 \\ 2, & 2 \leq x < 3 \\ \vdots & \end{cases}$$

In words, the absolute value of x is the distance x is from the origin.

The greatest integer function returns the value of the greatest integer less than x
(or x if x is an integer).

11.2   Examples

EX 1. Find the domain and range of each function graphed in Figures 11.1-11.10.

Solution ☞ The Constant Function has a domain of $\mathbb{R}$ and a range of k.

The Linear, Cubic, and Cube Root Functions have a domain and range of $\mathbb{R}$.

The Quadratic, Square Root, and Absolute Value Functions have a domain of $\mathbb{R}$ and a range of $x \geq 0$.

The Reciprocal Function has a domain of $(-\infty, 0) \cup (0, \infty)$ and a range of $(-\infty, 0) \cup (0, \infty)$.

The Square of the Reciprocal Function has a domain of $(-\infty, 0) \cup (0, \infty)$ and a range of $y > 0$.

The Greatest Integer Function has a domain of $\mathbb{R}$ and a range of $\{y|y$ is an integer$\}$.

EX 2. Sketch the graph of

$$h(x) = \begin{cases} 2, & -6 < x < -3 \\ x^3, & -3 \leq x < 0 \\ x, & 0 \leq x \leq 3 \\ x^2, & 3 < x \leq 4 \end{cases}$$

Solution ☞ This graph should be a horizontal line at $y = 2$ on the interval $(-6, -3)$, a cubic function on $[-3, 0)$, a linear function on $[0, 3]$ and a quadratic function on $(3, 4]$. Combining these portions together yields the following graph.

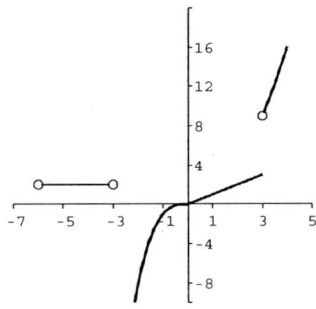

Figure 11.11: Graph of $y = h(x)$

EX 3. Evaluate $\lfloor -4.3 \rfloor$, $\lfloor -2 \rfloor$ and $\lfloor 1.2 \rfloor$.

Solution To determine the value of $\lfloor -4.3 \rfloor$, we will recall that this represents the greatest integer less than or equal to -4.3. This is -5, so $\lfloor -4.3 \rfloor = -5$.

Recognizing that -2 is an integer means that $\lfloor -2 \rfloor = -2$.

The greatest integer less than 1.2 is 1, so $\lfloor 1.2 \rfloor = 1$.

Name:_____

Section Number:_____

Date:_____

Chapter 11 Exercises

1. Evaluate $\lfloor 3 \rfloor$, $\lfloor 1.6 \rfloor$ and $\lfloor -2.4 \rfloor$.

$\lfloor 3 \rfloor = $ _____ $\lfloor 1.6 \rfloor = $ _____ $\lfloor -2.4 \rfloor = $ _____

For problems 2 - 3, sketch the graphs of the given piecewise functions.

2.   3.

$$g(x) = \begin{cases} x, & -4 \le x < -1 \\ \lfloor x \rfloor, & -1 \le x < 2 \\ x^3, & 2 \le x \le 4 \end{cases}$$

$$h(x) = \begin{cases} \sqrt[3]{x}, & -8 < x < -1 \\ x^3, & -1 \le x \le 1 \\ x, & 1 < x < 4 \end{cases}$$

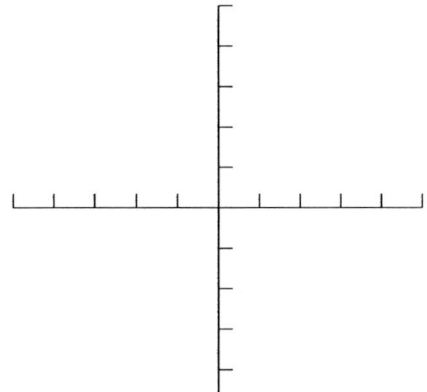

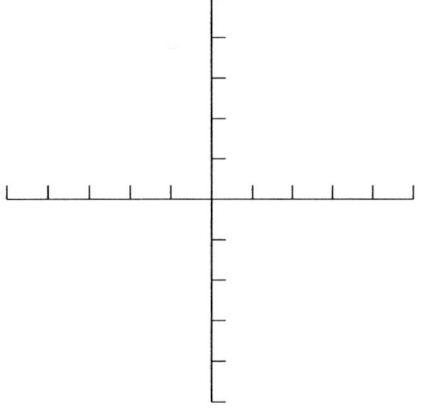

4. Which of the graphs in Figures 11.1-11.10 are even?

5. Which of the graphs in Figures 11.1-11.10 are odd?

For numbers 6 - 9, match the piecewise functions with their graphs. Record the letter of the correct answer in the space provided.

6.

$$f(x) = \begin{cases} \frac{1}{x^2}, & -\infty < x < 0 \\ \sqrt{x}, & 0 \le x < \infty \end{cases}$$

Answer:_____

7.

$$g(x) = \begin{cases} x^3, & -\infty < x < 0 \\ 1/x, & 0 \le x < \infty \end{cases}$$

Answer:_____

8.

$$h(x) = \begin{cases} x, & -\infty < x < 0 \\ |x|, & 0 \le x < \infty \end{cases}$$

Answer:_____

9.

$$k(x) = \begin{cases} \lfloor x \rfloor, & -\infty < x < 0 \\ \sqrt[3]{x}, & 0 \le x < \infty \end{cases}$$

Answer:_____

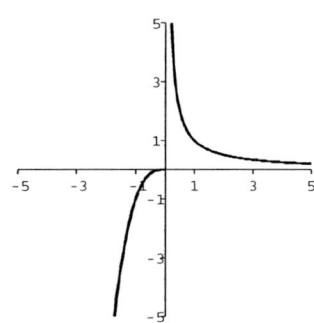

A.

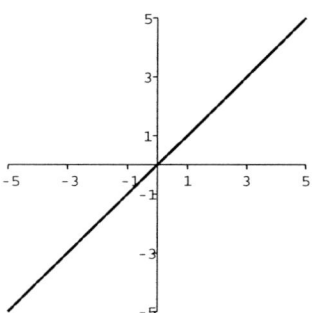

B.

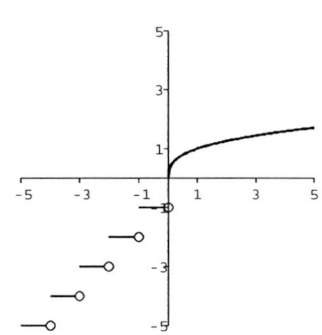

C.

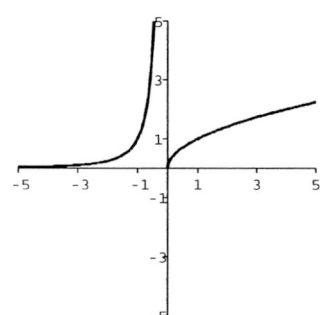

D.

106

For number 10, show all work and print the answer in the space provided.

10. Which of the following graphs does not use any portion of a $y = x^3$ graph?

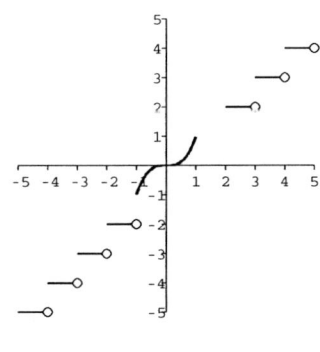

A.)

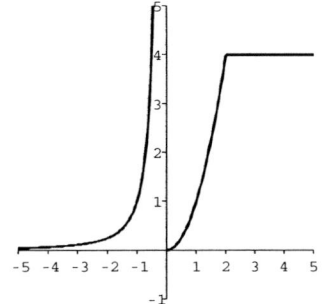

B.)

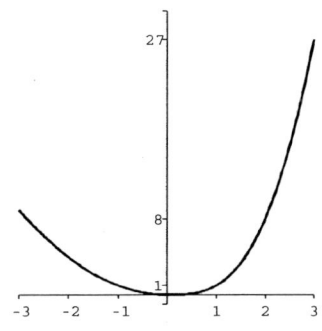

C.)

D.)

E.) None of these.

Answer:_____

107

Chapter 12

Shifting, reflecting and stretching graphs

12.1 Notes about Shifting, Reflecting and Stretching graphs

The table below shows the effect of some translations on the graph of $y = f(x)$ (assume that $a > 1$, $0 < b < 1$ and $c > 0$).

The graph of $y =$	will be the graph of $y = f(x)$
$f(x) + c$	shifted up c units
$f(x) - c$	shifted down c units
$af(x)$	vertically stretched by a factor of a
$bf(x)$	vertically shrunk by a factor of b
$-f(x)$	reflected across the x-axis
$f(x + c)$	shifted left c units
$f(x - c)$	shifted right c units
$f(ax)$	horizontally shrunk by a factor of $1/a$
$f(bx)$	horizontally stretched by a factor of $1/b$
$f(-x)$	reflected across the y-axis

12.2 ❖ Errors to Avoid

Be aware that the order in which the shifts, stretches, and reflections are performed in does matter, in much the same way that the order of operations for exponentiation, multiplication, etc. matters. One method for performing all of the translations correctly involves doing the translations in the following order: all reflections first, all stretching second and then all shifting. Within each step, the order does not matter; if there is both a vertical and horizontal stretch, the order in which they are performed does not matter.

For instance, when graphing $y = -(x - 3)^2 + 2$ the correct graph is not obtained by moving the graph up 2 units, right 3 units and then reflecting across the x-axis. The reflection should be done first and then the shifts.

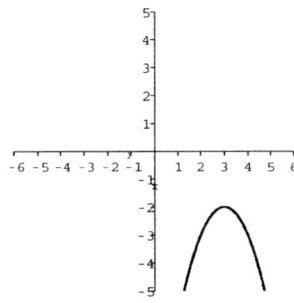

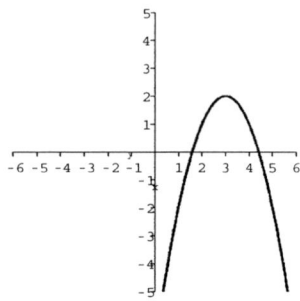

Figure 12.1: Wrong graph.   Figure 12.2: Right graph.

Further, sometimes the function you are asked to graph is not already in a 'usable' form. For instance, the rules do not easily apply to $g(x) = (5 - x)^3 + 4$. For a solution of how to graph this, see Example 3. See also Example 8.

12.3   Examples

EX 1. ★ (S03#5) The graph of the function $y = f(x - 1) - 3$ can be obtained by shifting the graph of $y = f(x)$:

A.) One unit to the left and 3 units downward.
B.) One unit to the right and 3 units downward.
C.) 3 units to the left and 1 unit downward.
D.) 3 units to the right and 1 units upward.
E.)   None of these.

Solution ☞ Here the graph is being shifted 1 unit to the right by the $(x - 1)$-term and 3 units down by the -3-term. Thus, the answer is B.).

EX 2. Sketch the graph of $y = 2x + 1$.

> **Solution** ☞ To sketch this graph, we will first notice that the underlying or *baseline* graph that is being translated is $y = x$. Once we recognize this, all that is necessary to do is apply the translations in the appropriate order. Here the graph of $y = x$ should be vertically stretched by a factor of 2 and then shifted up 1 unit.

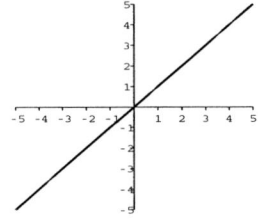 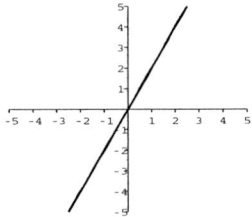 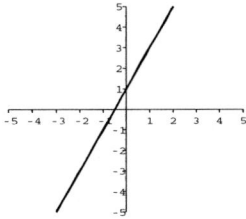

Figure 12.3: $y = x$.   Figure 12.4: $y = 2x$.   Figure 12.5: $y = 2x + 1$.

> The graph in the center is the graph of $y = x$ stretched vertically by a factor of 2. The graph at right is the vertically stretched graph shifted up 1 unit.

EX 3. Sketch the graph of $y = (5 - x)^3 + 4$.

> **Solution** ☞ To solve this problem, it is best to rewrite the original problem to better see how to apply the transformations. We notice that $f(x) = x^3$ is our baseline graph and that $y = (5 - x)^3 + 4 = (-(x - 5))^3 + 4$. Thus, $g(x)$ is the graph of x^3 reflected across the y-axis, moved right 5 and up 4. The the graph of $y = x^3$ and final graph are below.

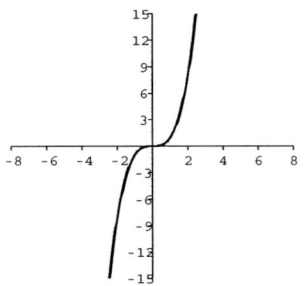 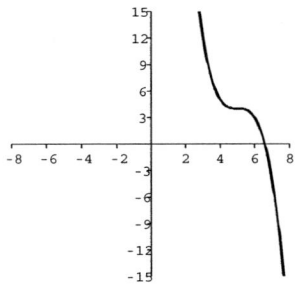

Figure 12.6: $y = x^3$.   Figure 12.7: $y = (5 - x)^3 + 4$.

EX 4. Sketch the graph of $y = 3\lfloor 2x \rfloor - 1$.

Solution ☞ Again it is important to recognize the baseline function here. In this case, it is $y = \lfloor x \rfloor$. Once this is recognized, it is only necessary to shrink the graph horizontally by a factor of $1/2$, stretch it vertically by a factor of 3 and shift the graph down 1 unit. This translations are shown below.

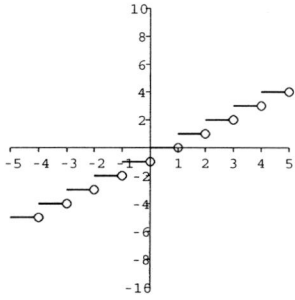

Figure 12.8: $y = \lfloor x \rfloor$.

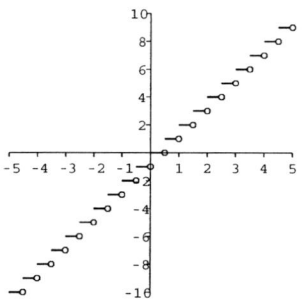

Figure 12.9: $y = \lfloor 2x \rfloor$.

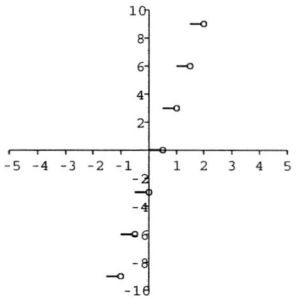

Figure 12.10: $y = 3\lfloor 2x \rfloor$.

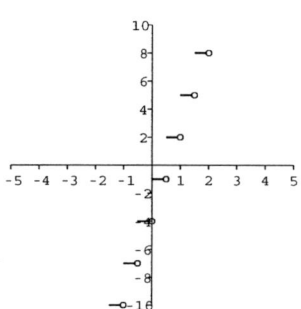

Figure 12.11: $y = 3\lfloor 2x \rfloor - 1$.

EX 5. ✭ (S04#19) The graph of $y = h(x)$ was obtained by transforming the graph of $y = |x|$. Find $h(x)$.

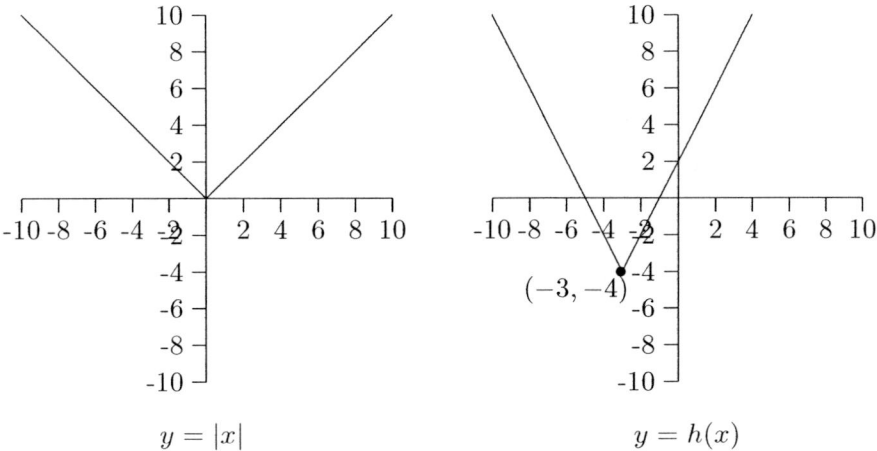

$$y = |x| \qquad\qquad y = h(x)$$

Solution ☞ Looking at the graph of $y = h(x)$, you will notice that the graph of $y = |x|$ has been shifted to the left 3 units and down 4 units. You should also notice that the graph is 'thinner'. This can be taken two ways: either the graph is stretched vertically by a factor of 2 or shrunk horizontally by a factor of 1/2. Thus, either $h(x) = 2|x+3| - 4$ or $h(x) = |2(x+3)| - 4$.

EX 6. ⋆ (F02#18c) The graph of $y = f(x)$ is given. Sketch the graph of $y = 2f(-x)$ on the axes provided. Include at least two ordered pairs on your final graph.

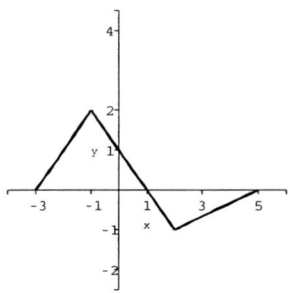

Figure 12.12: $y = f(x)$.

Solution ☞ The graph of $y = 2f(-x)$ is the graph of $y = f(x)$ reflected across the y-axis and stretched vertically by a factor of 2. Looking at this one point at a time, the point $(2, -1)$ is reflected to $(-2, -1)$ and stretched to $(-2, -2)$. Similarly, the points $(-3, 0)$, $(-1, -2)$, $(0, 1)$ and $(5, 0)$ can be translated to $(3, 0)$, $(1, 4)$, $(0, 2)$ and $(-5, 0)$, respectively.

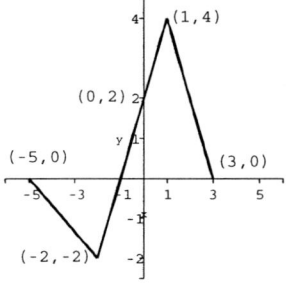

Figure 12.13: $y = 2f(-x)$.

EX 7. Where will the point $(1, 2)$ on the graph of $y = f(x)$ be shifted to on the graph of $y = -3f\left(\left(\frac{1}{4}\right)x\right) + 7$?

Solution ☞ Moving a single point works in a similar fashion to moving an entire graph. The figure below shows each of the following translations: reflect across the x-axis $(a.)$, vertically stretch by a factor of 3 $(b.)$, horizontally stretched by a factor of 4 $(c.)$ and then shifted up 7 units to the point $(4, 1)$.

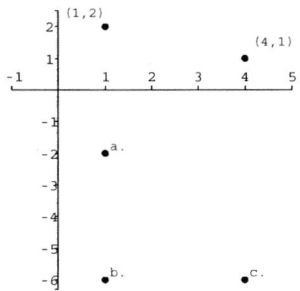

✎ Note that $-3f\left(\left(\frac{1}{4}\right)4\right) + 7 = -3f(1) + 7$. Because $f(1) = 2$, $-3f(1) + 7 = -3(2) + 7 = -6 + 7 = 1$.

EX 8. Where will the point $(1, -3)$ on the graph of $y = f(x)$ be shifted to on the graph of $y = -6f(15 - 3x) + 2$?

Solution ☞ As was done in the previous example, the shifting of the point $(1, -3)$ will be treated in the same manner the entire graph would. However, the function is not in an easily translatable form, so first note that $y = -6f(15 - 3x) + 2 = -6f(-3(x - 5)) + 2$. This form makes it easier to apply the translation rules.

The point should be:

- reflected across the y-axis to $(-1, -3)$,
- reflected across the x-axis to $(-1, 3)$,
- horizontally shrunk by a factor of 3 to $(-1/3, 3)$,
- vertically stretched by a factor of 6 to $(-1/3, 18)$,
- moved right 5 to $(14/3, 18)$ and up 2 to $(14/3, 20)$.

Thus, the point $(1, -3)$ on the graph of $y = f(x)$ will be shifted to $(14/3, 20)$ on the graph of $y = -6f(15 - 3x) + 2$

Alternate Solution ☞ Here, $15 - 3x = 1 \Rightarrow x = 14/3$ and $-6(-3) + 2 = 20$.

Name:_____

Section Number:_____

Date:_____

Chapter 12 Exercises

Determine the appropriate baseline function on problems 1 - 6 .

1. $y = 2(x-3)^2 + 5$.

Answer:_____

2. $y = 5 - |x + 2|$.

Answer:_____

3. $y = \lfloor x - 3 \rfloor + 2$.

Answer:_____

4. $y = -\sqrt{-x}$.

Answer:_____

5. $y = -(-x)^3$.

Answer:_____

6. $y = \dfrac{1}{(x-1)^2} + 3$.

Answer:_____

For problems **7 - 12** sketch the graphs of the functions described in **1 - 6**. Include the x- and y-intercepts.

7. $y = 2(x-3)^2 + 5$

8. $y = 5 - |x + 2|$

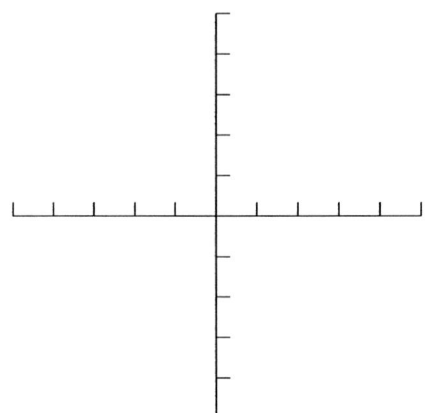 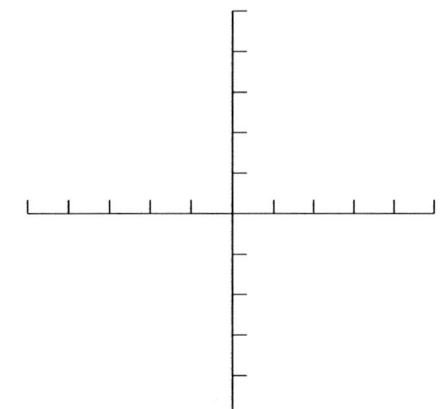

117

9. $y = \lfloor x - 3 \rfloor + 2$

10. $y = -\sqrt{-x}$

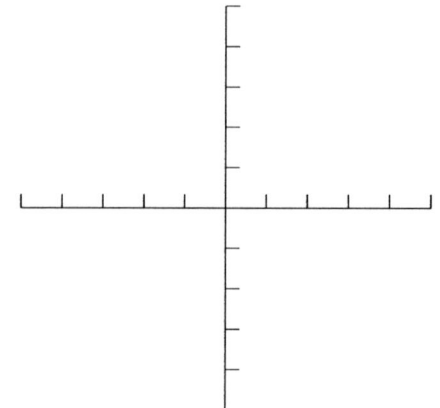

11. $y = -(-x)^3$

12. $y = \dfrac{1}{(x-1)^2} + 3$

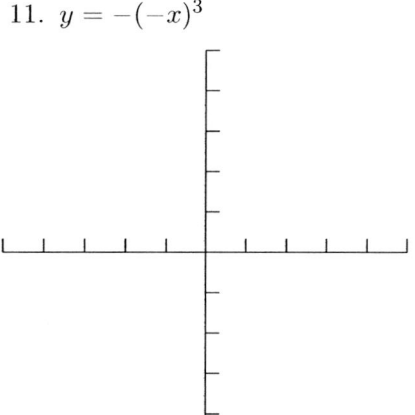

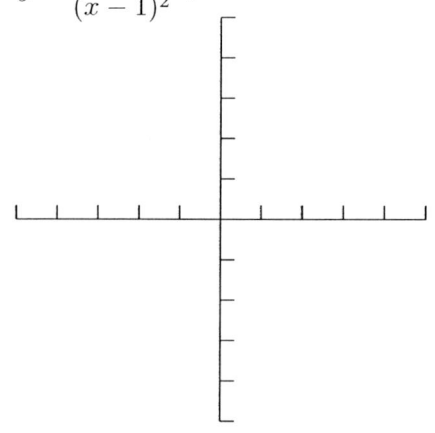

13. One of the graphs in problems **7 - 12** was identical to the corresponding answer in **1 - 6**. Which problem was that? What about the function made this the case?

14. Where will the point $(2, 4)$ on the graph of $y = f(x)$ be shifted to on the graph of $y = 3f(8 - 2x) + 1$?

15. Where will the point $(9, -2)$ on the graph of $y = h(x)$ be shifted to on the graph of $y = 2h(3x - 6) - 7$?

16. Given the graph of $y = f(x)$ below, sketch the graph of $y = -2f(x + 3) - 1$ in the space provided.

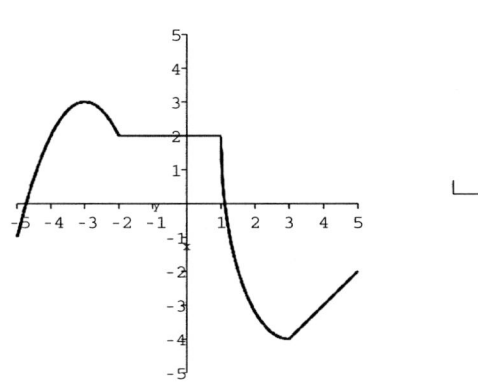

Figure 12.14: $y = f(x)$.

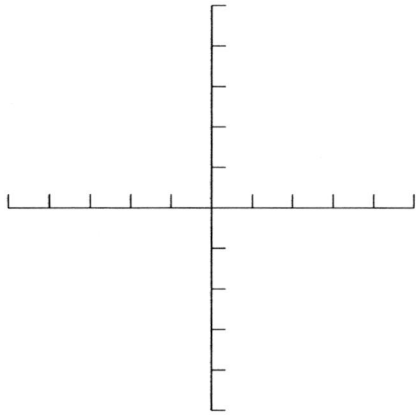

Figure 12.15: $y = -2f(x + 3) - 1$

For numbers 17 - 20, show all work and print the answer in the space provided.

17. Which of the following graphs represents $y = 2f(x) - 3$ if the graph of $y = f(x)$ is given?

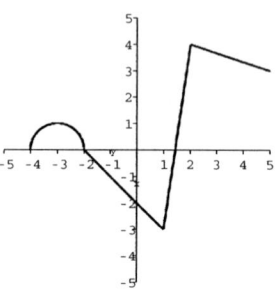

Figure 12.16: $y = f(x)$.

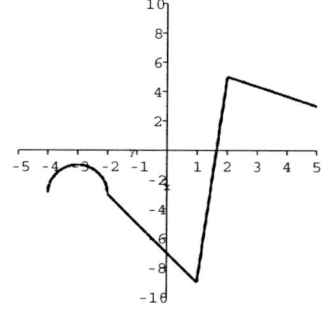

A.)

B.)

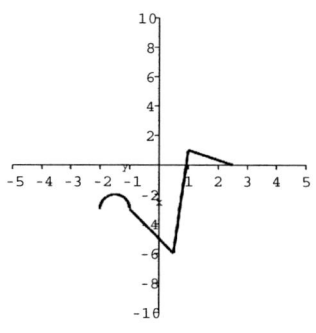

C.)

D.)

E.) None of these.

Answer:_____

18. Which of the following graphs represents $y = -g(3x) + 4$ if the graph of $y = g(x)$ is given?

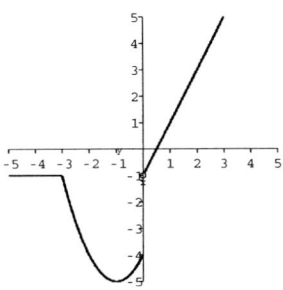

Figure 12.17: $y = g(x)$.

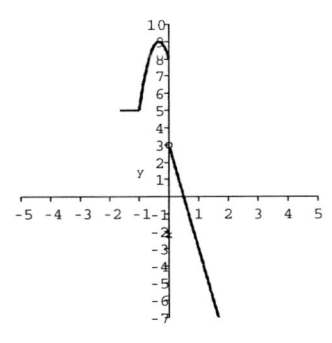

A.)

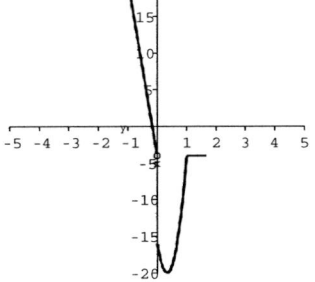

B.)

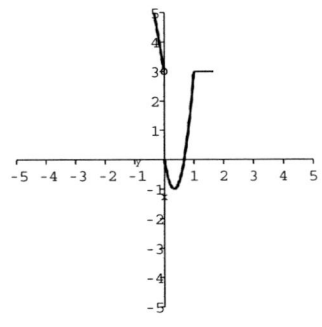

C.)

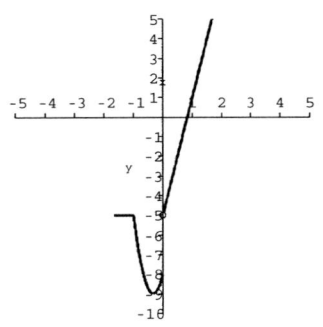

D.)

E.) None of these.

Answer:_____

19. Which of the following sets of directions will yield the correct graph of $y = -f(x + 1) - 2$ by shifting $y = f(x)$ in the order described?

    A.) Reflect the graph of f across the x-axis, move down 2 and left 1.
    B.) Move the graph of f down 2, left 1 and reflect the graph across the y-axis.
    C.) Reflect the graph of f across the y-axis, move up 2 and right 1.
    D.) Reflect the graph of f across the x-axis, move down 2 and right 1.
    E.) None of these.

Answer:_____

20. Which of the following sets of directions will yield the correct graph of $y = 5f(-2x + 4)$ by shifting $y = f(x)$ in the order described?

    A.) The graph of f is reflected across the x axis, horizontally stretched by a factor of 2, vertically stretched by a factor of 5 and shifted left 4.
    B.) The graph of f is reflected across the y axis, horizontally shrunk by a factor of 1/2, vertically stretched by a factor of 5 and shifted right 2.
    C.) The graph of f is reflected across the y axis, horizontally shrunk by a factor of 1/2, vertically stretched by a factor of 5 and shifted left 4.
    D.) The graph of f is reflected across the x axis, horizontally shrunk by a factor of 1/2, vertically stretched by a factor of 5 and shifted right 2.
    E.) None of these.

Answer:_____

Chapter 13

Combinations of functions

13.1   Notes about Combinations of functions

Below are five common combinations of functions and their definitions.

1. $(f + g)(x) = f(x) + g(x)$

   Addition of two functions.

2. $(f - g)(x) = f(x) - g(x)$

   Subtraction of two functions.

3. $(fg)(x) = f(x)g(x)$

   Multiplication of two functions.

4. $(f/g)(x) = \dfrac{f(x)}{g(x)}$

   Division of two functions.

5. $(f \circ g)(x) = f(g(x))$

   Combination of two functions.

In the first three combinations, the domain of the combination is the overlap of the domains of f and g. In the fourth, the domain is the overlap with the restriction that $g(x) \neq 0$. In the fifth, the domain is the set of x-values of the domain of $g(x)$ for which $g(x)$ is in the domain of $f(x)$.

13.2   ❖ Errors to Avoid

Note that $(f + g)(x) \neq (f(x) + g(x)) \cdot (x)$. Also, note that $(f \circ g)(x) \neq f(x) \cdot g(x)$.

13.3 Examples

EX 1. For $f(x) = 2x - 12$, $g(x) = x^3 + 8$ and $h(x) = \sqrt{x + 19}$, simplify the following combinations and find the domain of the combination:

(a) $(f + g)(x)$.

Solution ☞ $(f+g)(x) = f(x)+g(x) = (2x-12)+(x^3+8) = x^3+2x-4$. The domain of $f(x)$ is all real numbers and the domain $g(x)$ is all real numbers. Because the union of $\mathbb{R}$ and $\mathbb{R}$ is $\mathbb{R}$, the domain of $(f+g)(x)$ is all real numbers.

(b) $(h/f)(x)$.

Solution ☞ $(h/f)(x) = \dfrac{h(x)}{f(x)} = \dfrac{\sqrt{x + 19}}{2x - 12}$. The domain of $h(x)$ is $x \geq -19$ or $[-19, \infty)$, the domain of $f(x)$ is all real numbers, and $f(x) = 0$ when $x = 6$ so the domain of $(h/f)(x)$ is $[-19, 6) \cup (6, \infty)$.

(c) $(hg)(x)$.

Solution ☞ $(hg)(x) = h(x)g(x) = (\sqrt{x + 19})(x^3 + 8)$. The domain of $g(x)$ is all real numbers and the domain of $h(x)$ is $x \geq -19$ or $[-19, \infty)$ so the domain of $(hg)(x)$ is $[-19, \infty)$.

(d) $(f \circ g)(x)$.

Solution ☞ $(f \circ g)(x) = f(g(x)) = f(x^3+8) = 2(x^3+8)-12 = 2x^3+4$. The domain of $g(x)$ is all real numbers and the domain of $f(x)$ is all real numbers, so the domain of $(f \circ g)(x)$ is also all real numbers.

(e) $(g \circ f)(x)$.

Solution ☞ $(g \circ f)(x) = g(f(x)) = g(2x - 12) = (2x - 12)^3 + 8$. The domain of $f(x)$ is all real numbers and the domain of $g(x)$ is all real numbers, so the domain of $(g \circ f)(x)$ is also all real numbers.

(f) $(h \circ g)(x)$.

Solution ☞ $(h \circ g)(x) = h(g(x)) = h(x^3 + 8) = \sqrt{(x^3 + 8) + 19} = \sqrt{x^3 + 27}$. The domain of $g(x)$ is all real numbers. The domain of $h(x)$ is $x \geq -19$, so we must determine where $g(x) \geq -19$.

$$g(x) \geq -19 \Rightarrow x^3 + 8 \geq -19 \Rightarrow$$
$$x^3 + 27 \geq 0 \Rightarrow (x + 3)(x^2 - 3x + 9) \geq 0 \Rightarrow$$
$$x + 3 \geq 0 \Rightarrow x \geq -3.$$

So, the domain of $(h \circ g)(x)$ is $x \geq -3$.

EX 2. Use the graphs of $y = f(x)$ and $y = g(x)$ below to answer the questions that follow.

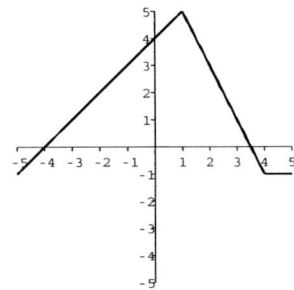

Figure 13.1: $y = f(x)$ Figure 13.2: $y = g(x)$

(a) Evaluate $g(f(1))$.

   Solution ☞ We see that $f(1) = 5$. Thus, $g(f(1)) = g(5)$ and $g(5) = -2$.

(b) Graph $y = (f - g)(x)$.

   Solution ☞ The graph on the left below shows several points that could be found by plugging in x-values. For instance, $(f - g)(-5) = f(-5) - g(-5) = -1 - (-6) = 5$ meaning there is a point $(-5, 5)$ on the final graph. The points chosen are the endpoints of the graph and the points at which at least one of the graphs changed in some way (e.g. increasing to decreasing, etc.).

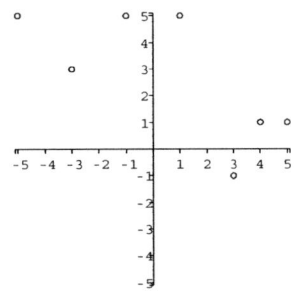
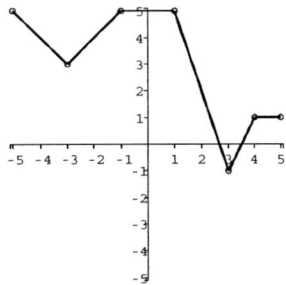

Figure 13.3: Some points on the graph of $(f - g)(x)$

Figure 13.4: $y = (f - g)(x)$

(c) For what values of x will $(g/f)(x)$ be undefined?

   Solution ☞ $(g/f)(x)$ will be undefined when $g(x)$ is undefined, $f(x)$ is undefined, or $f(x) = 0$. Because $f(x)$ and $g(x)$ are always defined, we need only determine where $f(x) = 0$, which is at $x = -3, 3$, and 5.

EX 3. Given $f(x) = \sqrt{2x - 8}$ and $g(x) = \sqrt{9 - x}$, find the domains of

(a) $(f \circ g)(x)$.

Solution ☞ First we need to find the domain of $f(x)$ and $g(x)$. The domain of $f(x)$ is $x \geq 4$ and the domain of $g(x)$ is $x \leq 9$. To find the domain of $f(g(x))$, we need to determine the values of x in the domain of $g(x)$ for which $g(x)$ is in the domain of $f(x)$. We know that the x-values are limited by $x \leq 9$ and we need to determine when $g(x) \geq 4$. This means $\sqrt{9 - x} \geq 4$ or equivalently $9 - x \geq 16$. Solving for x yields $x \leq -7$. Thus, we must determine the values where $x \leq 9$ and $x \leq -7$ overlap, which is $x \leq -7$.

(b) $(g \circ f)(x)$.

Solution ☞ Knowing the domains of $f(x)$ and $g(x)$ we need only to determine the values of x in the domain of $f(x)$ for which $f(x)$ is in the domain of $g(x)$. So we know that $x \geq 4$ and we need to determine when $f(x) \leq 9$ or $\sqrt{2x - 8} \leq 9$ or $2x - 8 \leq 81$. Solving for x yields $x \leq 44.5$. Determining where $x \geq 4$ and $x \leq 44.5$ overlap yields $4 \leq x \leq 44.5$.

EX 4. Find functions $f(x)$ and $g(x)$ so that $h(x) = f(g(x))$.

(a) $h(x) = (x + 1)^2$.

Solution ☞ To determine an answer, it can be helpful to think of $g(x)$ as the 'inside' function. Using this idea, we see that $g(x)$ could be $x + 1$ is one possible answer. If $g(x) = x + 1$, then $f(x)$ must be x^2.

Note this solution is not unique; another possible answer could be $f(x) = (x - 1)^2$ and $g(x) = x + 2$. Check it and see.

(b) $h(x) = (7x + 2)^3 - 6(7x + 2) + 9$

Solution ☞ Again thinking of $g(x)$ as the inside function allows us to see $g(x)$ could be $7x + 2$. This would leave $f(x) = x^3 - 6x + 9$.

(c) $h(x) = \sqrt{5 - x} + 7(x - 5) + 3$

Solution ☞ A possible value of $g(x)$ is $5 - x$. This means that $f(x) = \sqrt{x} + 7(-x) + 3$. Alternatively, if you let $g(x) = x - 5$ then $f(x)$ would be $\sqrt{-x} + 7x + 3$.

Chapter 13 Exercises

Use $f(x) = 4x - 7$, $g(x) = (x - 2)^3 + 4$ and $h(x) = 5x + \sqrt{x - 3}$ to answer questions 1 - 5.

1. Find $(g - f)(x)$ and its domain.

2. Find $(hf)(x)$ and its domain.

3. Find $(h/g)(x)$ and its domain.(*Hint:* Sketch the graph of $y = g(x)$)

4. Find $h(f(x))$ and its domain.

5. Find $g(f(h(x)))$.

For numbers 6 - 10, find the indicated value using the graphs below.

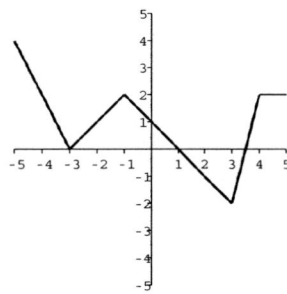

Figure 13.5: $y = f(x)$

Figure 13.6: $y = g(x)$

6. $(f + g)(3) =$

7. $(g/f)(-1) =$

8. $f(g(3)) =$

9. $g(f(5)) =$

10. $f(f(f(1))) =$

Determine $f(x)$ and $g(x)$ if $h(x) = g(f(x))$ for numbers 11 - 14 (don't use $f(x) = x$).

11. $h(x) = (5 - x)^3$.

$f(x) = $ _____

$g(x) = $ _____

12. $h(x) = (x + x^2)^3 - 5(x + x^2) + 6$.

$f(x) = $ _____

$g(x) = $ _____

13. $h(x) = \sqrt{10x + 3} - \dfrac{1}{10x + 3} + 10x$.

$f(x) = $ _____

$g(x) = $ _____

14. $h(x) = x - 1 + (x - 1)^3 - \sqrt{x + 1}$.

$f(x) = $ _____

$g(x) = $ _____

For numbers 15 - 18, show all work and print the answer in the space provided.

15. If $f(x) = x^2 - 3$ and $g(x) = 15 - 2x - x^2$, find $(f - g)(x)$.

    A.) $2x^2 + 2x - 12$ B.) $(15 - 2x - x^2)^2 - 3$

    C.) $15 - 2(x^2 - 3) - (x^2 - 3)^2$ D.) $12 + 2x$

    E.) None of these.

Answer:_____

16. Let $f(x) = \sqrt{x}$ and $g(x) = x^2 + 5$. Find $(f \circ g)(2)$.

    A.) 7   B.) 4   C.) -1 D.) $9\sqrt{2}$ E.) 3

Answer:_____

17. If $f(1) = 3$, $f(3) = 4$, $f(4) = 7$, $g(1) = 4$, $g(3) = 5$ and $g(4) = 2$, what is $g(f(1))$?

    A.) 7   B.) 4   C.) 2   D.) 5   E.) 1

Answer:_____

18. The graphs of $y = f(x)$ and $y = g(x)$ are below.

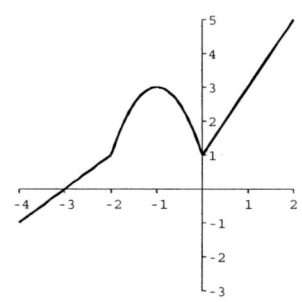

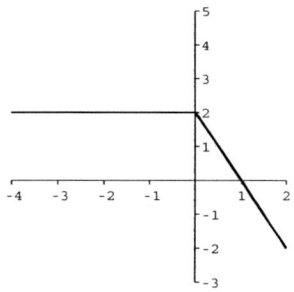

Figure 13.7: $y = f(x)$ Figure 13.8: $y = g(x)$

Which of the graphs below is the correct graph of $y = (f + g)(x)$?

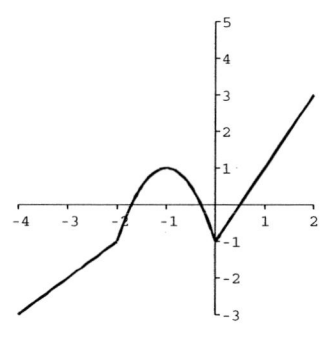

A.)

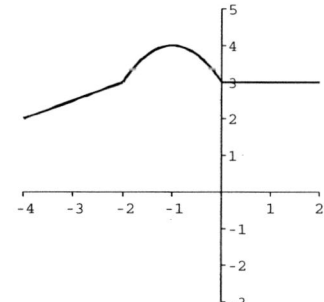

B.)

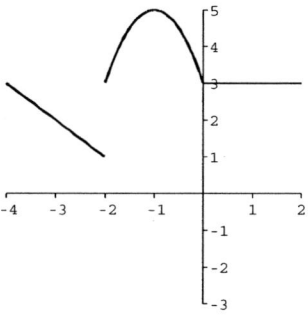

C.)

D.)

E.) None of these.

Answer:_____

131

Chapter 14

Inverse functions

14.1   Notes about Inverse functions

Definition: An *inverse function*, g, for the function f is a function such that $f(a) = b \iff g(b) = a$. This means that the domain of f must coincide with the range of g and the range of f must be the domain of g.

A common notation for the inverse of f is f^{-1}. f^{-1} is read "f inverse", not "f to the negative first."

If f has an inverse, then $f(f^{-1}(x)) = x$ and $f^{-1}(f(x)) = x$ on the appropriate domain. The domain of f^{-1} is the range of f and the range of f^{-1} is the domain of f.

To graph the inverse of a function, reflect the graph the original function across the line $y = x$. This has the effect of exchanging the x-values for the y-values and vice versa (which is exactly the operation performed by finding an inverse).

Definition Only certain types of functions have inverses that are also functions. A necessary condition for a function to have an inverse is that it must be *one-to-one*. The graph of any one-to-one function will pass the *horizontal line test*. The horizontal line test states that for a function to be one-to-one, there can be no horizontal line that crosses the graph more than once. This is because if a graph that was crossed more than once horizontally is reflected across the line $y = x$, it will fail the vertical line test and won't be a function.

14.2   Examples

EX 1. Determine if the following functions have inverses:

(a) $f(x) = 2x - 3$.

Solution ☞ To see if f has an inverse, we will graph the function to see if it passes the horizontal line test. The graph is below.

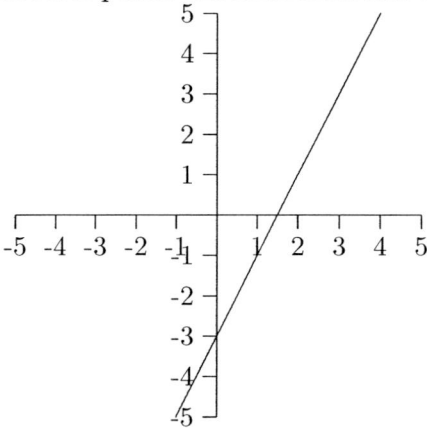

Because f passes the h.l.t., f has an inverse.

(b) $g(x) = x^2 - 5$.

Solution ☞ To see if G has an inverse, we will graph the function to see if it passes the horizontal line test. The graph is below.

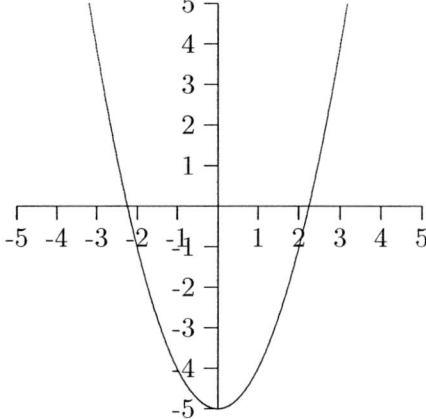

Because g fails the h.l.t., g does not have an inverse.

However, if we only look at the portion of the graph to the right 0, we see that that portion by itself would pass the h.l.t. So the function $g(x) = x^2 - 5$ for $x \geq 0$ DOES have an inverse. See also EX 3.

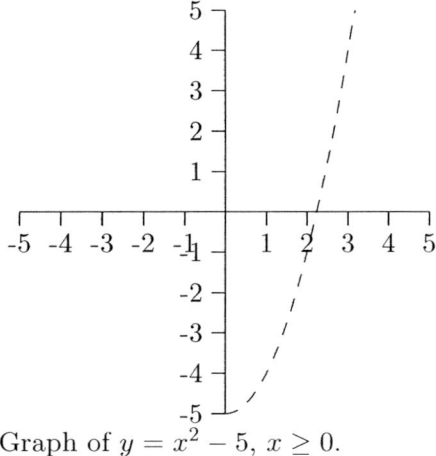

Graph of $y = x^2 - 5$, $x \geq 0$.

EX 2. Find the inverse of $f(x) = 2x - 3$.

Solution ☞ From EX 1, we know that f has an inverse. To find the inverse, we will exchange the x and y in $y = 2x - 3$ and then solve the new equation for y. Note that this method agrees with the original definition of the inverse.

$$y = 2x - 3 \Rightarrow$$
$$x = 2y - 3 \Rightarrow$$
$$x - 3 = 2y \Rightarrow$$
$$y = \frac{x - 3}{2}.$$

So, $f^{-1}(x) = \dfrac{x - 3}{2}$.

EX 3. Find the inverse of $g(x) = x^2 - 5$ on the domain $x \geq 0$.

Solution ☞ From EX 1b, we saw that when the domain of $g(x)$ is restricted to $x \geq 0$, then g does have an inverse.

The work to find the inverse is below:

$$y = x^2 - 5 \Rightarrow x = y^2 - 5 \Rightarrow$$

$$x - 5 = y^2 \Rightarrow \sqrt{x - 5} = y \Rightarrow g^{-1}(x) = \sqrt{x - 5}.$$

Note: Normally when taking the square root of both sides, there are two solutions. However, because of how we restricted the domain, only the positive solution remains.

EX 4. Sketch the graph of the inverse of $y = f(x)$ using the graph below.

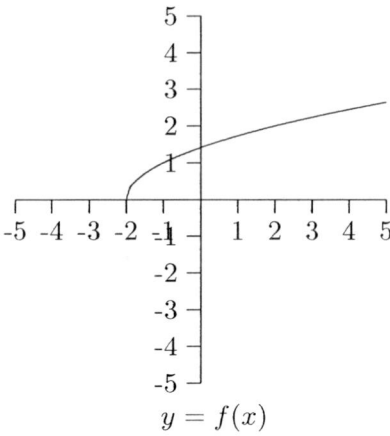

$$y = f(x)$$

Solution ☞ To find the graph of the inverse, we need to reflect the graph across the line $y = x$. The graph of $y = f(x)$ is shown below with the graph of $y = x$ shown as a dashed line.

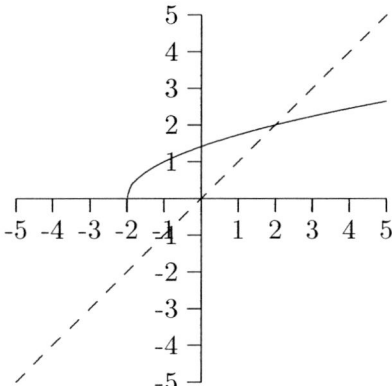

Since the points $(-2, 0)$ and $(2, 2)$ are on the graph of $y = f(x)$, then the points $(0, -2)$ and $(2, 2)$ are on the graph of $y = f^{-1}(x)$. Also, the part of the graph that is 'above' (up and to the left) of the line $y = x$ will get reflected 'below' the graph of $y = x$. Similarly, the part of the graph that is 'below' (down and to the right) of the line $y = x$ will get reflected 'above' the graph of $y = x$. The final result of this reflection is shown below.

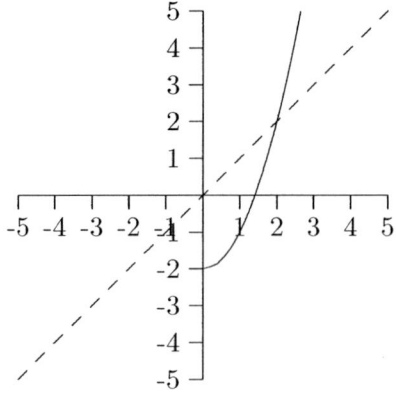

Chapter 14 Exercises

For problems 1 - 4, determine if the given function has an inverse over all real numbers.

   1. $f(x) = -5x + 6$.

   2. $g(x) = -\sqrt{3 - x}$.

   3. $f(x) = x^4 + 6$.

4. $g(x) = (x - 2)^3 + 1$.

For problems 5 - 9, determine if the given function has an inverse over all real numbers. If it does, then find the inverse. If it does not, then find a domain over which it does have an inverse.

5. $f(x) = -10x + 7$

6. $g(x) = 8x - 15$

7. $f(x) = \dfrac{-x-7}{2x+6}$

8. $g(x) = \dfrac{5-3x}{x+4}$

9. $f(x) = (x+3)^2 - 1$

For numbers 10-11, show all work and print the answer in the space provided.

10. Suppose that $f(1) = -2$, $f(3) = 7$ and $f^{-1}(-1) = 6$. Which of the following is true?

    A.) $f(-2) = 1$ B.) $f(6) = -1$ C.) $f^{-1}(7) = -3$

    D.) $f(-1) = 6$ E.) $f^{-1}(3) = 7$

Answer:_____

11. Use the graph of $y = f(x)$ below to determine which of the following is the graph of $y = f^{-1}(x)$.

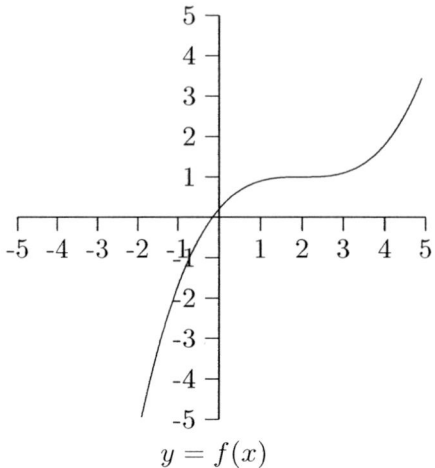

$$y = f(x)$$

A.)

B.)

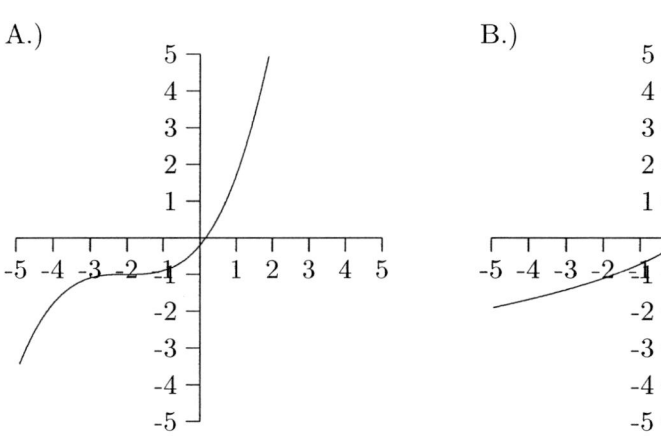

C.)

D.)

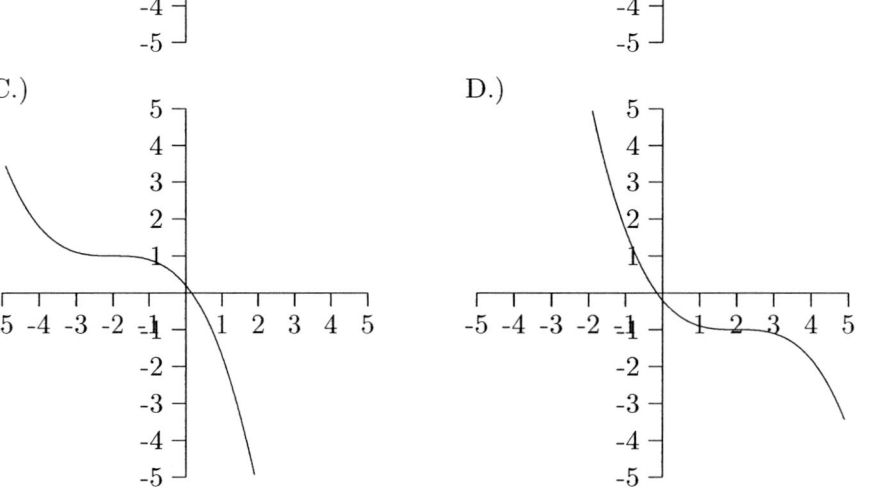

E.) $f(x)$ does not have an inverse because it is not one-to-one.

Answer:_____

Chapter 15

Circles

15.1 Notes about Circles

Definition: The **distance** between two points (x_1, y_1) and (x_2, y_2) is given by the formula $D = \sqrt{(x_2 - x_1)^2 + (y_2 - y_1)^2}$.

Definition: The **midpoint** between two points (x_1, y_1) and (x_2, y_2) is the point $\left(\dfrac{x_1 + x_2}{2}, \dfrac{y_1 + y_2}{2}\right)$.

Definition: The equation of any circle can be written in the *standard form*, which is given by $(x - h)^2 + (y - k)^2 = r^2$, where h represents the x-value of the center, k represents the y-value of the center, and r represents the radius, the distance each point on the circle is from the center.

15.2 ❖ Errors to Avoid

When using the standard form of a circle be careful about the signs of the $x-$ and $y-$values of the center. For instance, the center of the circle $(x-3)^2 + (y+2)^2 = 25$ is $(3, -2)$. Also be sure to find the square root of the r^2 term to find the radius. The radius of $(x - 3)^2 + (y + 2)^2 = 25$ is 5, not 25.

15.3   Examples

EX 1. Find the distance between the points $(-1, 2)$ and $(2, 7)$.

> **Solution** ☞ Using the distance formula, we see that the distance between the points is given by $D = \sqrt{(2 - -1)^2 + (7 - 2)^2} = \sqrt{3^2 + 5^2} = \sqrt{34}$.

EX 2. Find the midpoint of the line segment that connects the points $(1, 4)$ and $(-3, 5)$.

> **Solution** ☞ Using the midpoint formula, we see that $\left(\dfrac{1 + -3}{2}, \dfrac{4 + 5}{2}\right) = \left(-1, \dfrac{9}{2}\right)$ is the midpoint.

EX 3. Sketch the graph of the circle $(x - 5)^2 + (y + 2)^2 = 16$.

> **Solution** ☞ Recognizing the equation as the standard form of a circle, we see that $h = 5$, $k = -2$ and $r^2 = 16$. Because the radius represents a length, $r = 4$ only. Thus, this circle has a center at $(5, -2)$ and a radius of 4. Further, the circle must have a point 4 units directly above, below, left and right of $(5, -2)$. These points are at $(5, 2)$, $(5, -6)$, $(1, -2)$ and $(9, -2)$, respectively, and are plotted below, as is the completed circle.

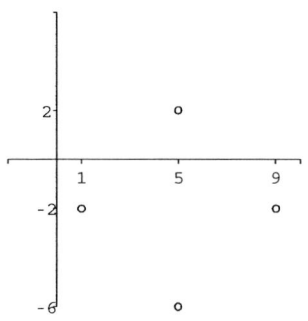
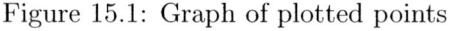

Figure 15.1: Graph of plotted points

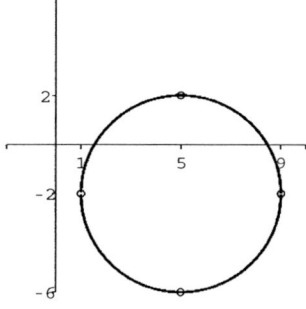

Figure 15.2: Graph of circle through plotted points

EX 4. ★ (S02#17) Consider a circle with center at $C = (2, -1)$ that goes through the point $(-5, 3)$.

(a) Find the radius of the circle.

Solution ☞ The radius of the circle is the distance from the center of the circle to any point on the outside of the circle. Using the distance formula gives:

$$R = \sqrt{(2 -\text{ } 5)^2 + (-1 - 3)^2} = \sqrt{7^2 + (-4)^2} = \sqrt{65}.$$

(b) Express the equation of this circle in standard form.

Solution ☞ We are told that $h = 2$ and $k = -1$, so $(x - 2)^2 + (y + 1)^2 = 65$.

EX 5. Suppose that the points $(-1, 2)$ and $(3, 2)$ are endpoints of a diameter of a circle. Find the equation of the circle in standard form.

Solution ☞ To find the equation of the circle, we need to find the center and the radius. The center is halfway between to endpoints of a diameter. Note that this is the midpoint of these two points. Thus the center is at $\left(\dfrac{-1+3}{2}, \dfrac{2+2}{2}\right) = (1, 2)$.

To find the radius, note that the radius is half the diameter of a circle. The diameter of this circle is the distance between the endpoints given which is $D = \sqrt{(3 - -1)^2 + (2 - 2)^2} = \sqrt{16 + 0} = 4$. Thus the radius is 2.

Putting this information together, we see that the circle is given by $(x - 1)^2 + (y - 2)^2 = 4$.

EX 6. ★ (S01#11) Find the center C and the radius R of the circle represented
by the equation $x^2 + y^2 - 4x + 10y + 13 = 0$.

A.) $C = (2, -5); R = 4$ B.) $C = (-2, 5); R = 16$ C.) $C = (2, -5); R = 16$

D.) $C = (4, -10); R = 4$ E.) $C = (-2, 5); R = 4$

Solution ☞ To solve this problem, we need to put the given form of the
circle in standard form. To do this will require completing the square twice.
The work for this is shown below.

$$x^2 + y^2 - 4x + 10y + 13 = 0 \Rightarrow x^2 - 4x + y^2 + 10y = -13$$

$$x^2 - 4x + 4 + y^2 + 10y + 25 = -13 + 4 + 25 \Rightarrow (x - 2)^2 + (y + 5)^2 = 16$$

Thus, the center is $(2, -5)$ and the radius is 4. The answer is **A**.

Chapter 15 Exercises

For problems 1 - 2, sketch the graph of the given circle on the graph provided.

1. $(x - 3)^2 + (y - 4)^2 = 9$.

2. $(x + 5)^2 + (y - 2)^2 = 16$.

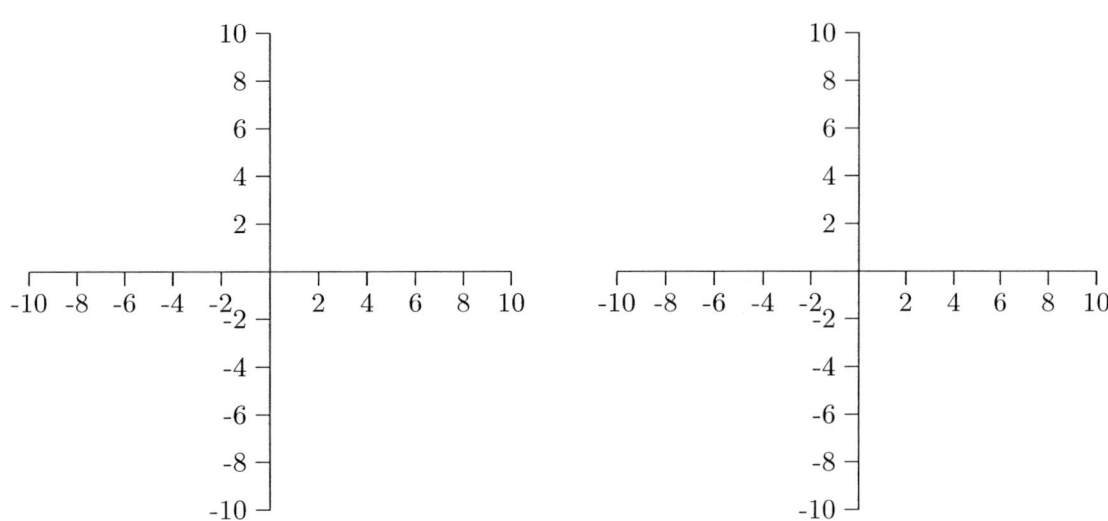

For problems 3 - 6, find the equation of the circle in standard form with the given information.

3. Center at $(-1, 4)$ and radius of 3.

4. Center at $(-2, -5)$ and radius of $\sqrt{7}$.

5. Center at $(3, -1)$ with a point on the circle at $(5, 4)$.

6. Endpoints of the diameter of a circle at $(-1, 5)$ and $(2, 1)$.

For problems 7 - 10, find the radius and center of the given circle and graph the circle on the axes provided.

7. $x^2 + y^2 = 25$

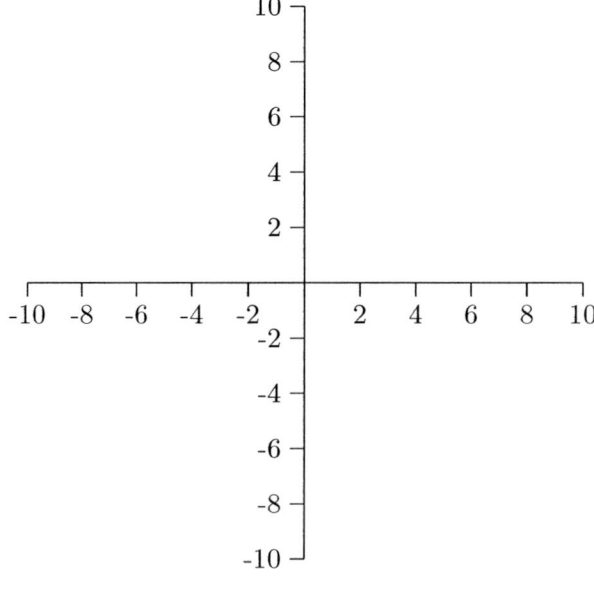

8. $x^2 - 4x + y^2 = 21$

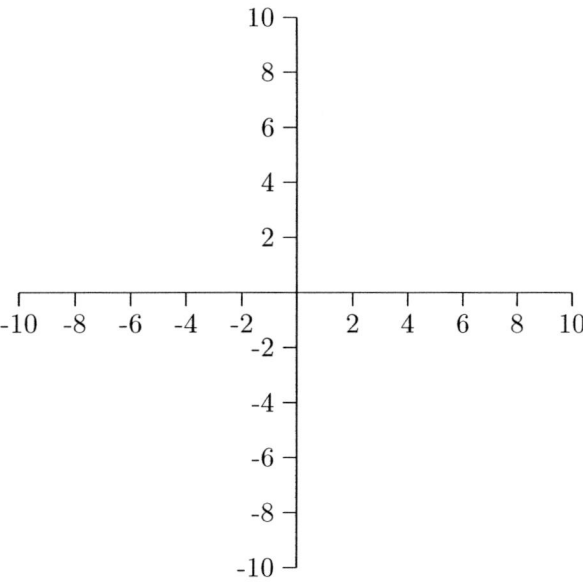

9. $x^2 + y^2 + 6y = 0$

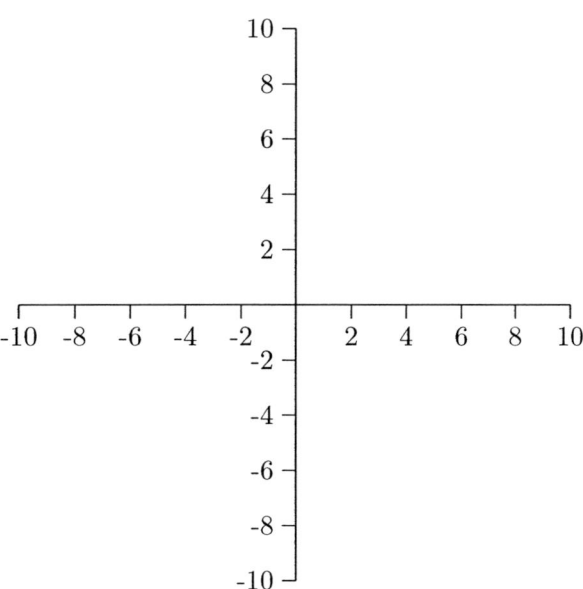

147

10. $x^2 + 10x + y^2 - 6y = -18$

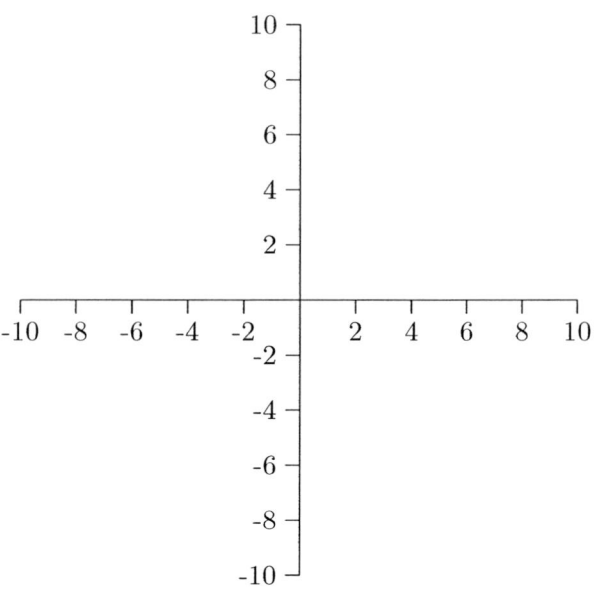

For number 11, show all work and print the answer in the space provided.

11. Find the center C and the radius R of the circle represented by the equation $x^2 + y^2 + 6x + 10y = 0$.

A.) $C = (6, 10); R = 0$ B.) $C = (-3, -5); R = 34$ C.) $C = (3, 5); R = 8$

D.) $C = (-3, -5); R = \sqrt{34}$ E.) $C = (-3, 5); R = 34$

Answer:_____

148

Chapter 16

Quadratic Functions

16.1 Notes about Quadratic Functions

Definition: Any function of the form $f(x) = ax^2 + bx + c$, where a, b, and c are real numbers and $a \neq 0$, is a *quadratic function*. The graph of a quadratic function is a *parabola*. If a is positive, the parabola will open upwards. If a is negative, the parabola will open downwards.

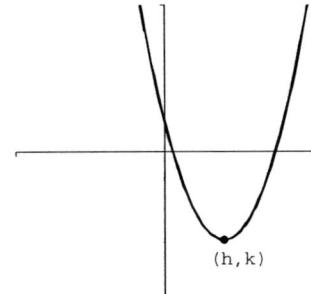

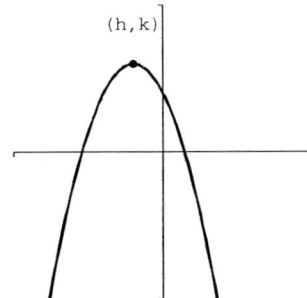

Figure 16.1: $a > 0$, Parabola opens up.   Figure 16.2: $a < 0$, Parabola opens down.

Definition: The *standard form* of a quadratic function is $f(x) = a(x - h)^2 + k$, where (h, k) is the *vertex* of the parabola (see Figures 16.1 and 16.2).

The y-value of the vertex will always be either the maximum $(a < 0)$ or minimum $(a > 0)$ value the function achieves.

If the quadratic function is in the $f(x) = ax^2 + bx + c$ form, then the vertex has an x-value of $\frac{-b}{2a}$ (the y-value is found by substituting the x-value into the function). To see this, we'll rewrite $f(x) = ax^2 + bx + c$ in standard form by completing the square.

$$
\begin{aligned}
f(x) &= ax^2 + bx + c \\
&= a(x^2 + (b/a)x) + c \\
&= a\left(x^2 + (b/a)x + \frac{b^2}{4a^2}\right) + c - \frac{b^2}{4a} \\
&= a\left(x + \frac{b}{2a}\right)^2 + c - \frac{b^2}{4a}.
\end{aligned}
$$

Here $h = \dfrac{-b}{2a}$, as expected.

16.2   Examples

EX 1. Determine the maximum and minimum value of the function $h(t) = 3t^2 - 6t - 2$.

Solution ☞ To find the maximum or minimum value of $h(t)$, we'll first find the vertex. The t-value of the vertex is $t = \frac{-(-6)}{2(3)} = 1$, so the minimum value is $h(1) = 3(1)^2 - 6(1) - 2 = -5$. We know this is a minimum value because $a > 0$ and thus the graph of h opens upward. Further, because the function continues upwards both to the left and the right of $t = 1$, h has no maximum value.

EX 2. ★ (S01#15) Consider the quadratic function $f(x) = x^2 + 6x + 7$.

a) Express $f(x)$ in standard form.

Solution ☞ Converting to standard form will require completing the square and is shown here: $f(x) = x^2 + 6x + 7 = (x^2 + 6x) + 7 = (x^2 + 6x + 9) + 7 - 9 = (x + 3)^2 - 2$.

b) State the coordinates of its vertex and the x and y intercepts.

   Solution ☞ To find the coordinates of the vertex, we'll use the x-value given by $\frac{-b}{2a}$ which is $\frac{-6}{2(1)}$ or -3. The y-value is $f(-3) = (-3)^2 + 6(-3) + 7 = -2$.

   To find the x-intercepts we'll set $f(x)$ equal to 0 and solve for x. Doing this using the quadratic equation yields $\dfrac{-6 \pm \sqrt{6^2 - 4 \cdot 1 \cdot 7}}{2} \approx -1.586, -4.414$.

   To find the y-intercept we'll substitute $x = 0$ into $f(x)$ yielding $f(0) = 0^2 + 6 \cdot 0 + 7 = 7$.

c) Sketch its graph.

   Solution ☞ To sketch the graph, first plot the vertex and the x- and y-intercepts.

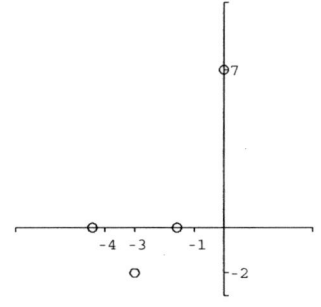   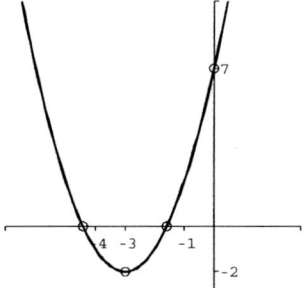

Figure 16.3: Plot of x- and y- inter-   Figure 16.4: $y = x^2 + 6x + 7$.
cepts and vertex.

EX 3. ✯ (S02#5) In which of the following is the standard form of the parabola $y = 3x^2 + 6x - 4$ given correctly?

A.) $y = 3(x + 1)^2 - 1$ B.) $y = (x + 1)^2 - 3$ C.) $y = 3(x + 3)^2 + 20$

D.) $y = 3(x - 1)^2 - 7$ E.) $y = 3(x + 1)^2 - 7$

   Solution ☞ To find the standard form of $y = 3x^2 + 6x - 4$ we will need to complete the square and this is shown here: $y = 3x^2 + 6x - 4 = 3(x^2 + 2x) - 4 = 3(x^2 + 2x + 1) - 4 - 3 = 3(x + 1)^2 - 7$ or E.).

EX 4. ★ (S03#19) Let $f(x) = -x^2 + 4x + 5$.

(a) Put $f(x)$ into the standard form of $f(x) = a(x - h)^2 + k$. Be sure to show the "complete the square" steps.

Solution ☞ As noted in the question, we will need to complete the square. The steps showing this are below:

$$\begin{aligned}
f(x) = -x^2 + 4x + 5 &= -(x^2 - 4x) + 5 \\
&= -(x^2 - 4x + 4) + 5 + 4 \\
&= -(x - 2)^2 + 9
\end{aligned}$$

Thus, $f(x) = -(x - 2)^2 + 9$ and the vertex is $(2, 9)$.

(b) Find the x- and y- intercepts.

Solution ☞ The x-intercepts can be found by solving $-x^2 + 4x + 5 = 0$. Factoring yields $-(x - 5)(x + 1) = 0$ which means $x = 5$ and $x = -1$ are the solutions and $(5, 0)$ and $(-1, 0)$ are the x-intercepts.

The y-intercept is found by plugging in 0 to the original function giving $f(0) = -(0)^2 + 4(0) + 5 = 5$ and thus the y-intercept is at $(0, 5)$.

(c) Graph the function on the axes below.

Solution ☞ The points known already (the vertex and intercepts) are plotted below as well as the final graph.

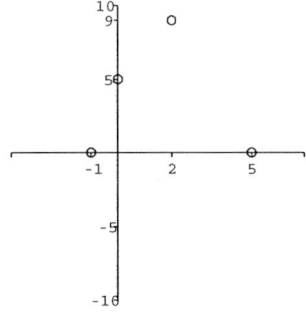

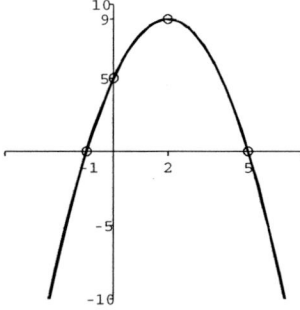

Figure 16.5: Plot of x- and y- ints. and vertex

Figure 16.6: $y = x^2 + 6x + 7$

Chapter 16 Exercises

On problems 1 - 6, find the vertex of the given quadratic equation.

1. $f(x) = (x-2)^2 + 3$

Answer:_____

2. $g(x) = 2(x+4)^2 + 5$

Answer:_____

3. $h(x) = (5-x)^2 - 7$

Answer:_____

4. $f(t) = t^2 + 2t + 1$

Answer:_____

5. $g(t) = -3t^2 + 1$

Answer:_____

6. $h(t) = 6t - t^2 + 5$

Answer:_____

For problems 7 - 10, put the given quadratic equation in standard form. Show all work.

7. $f(x) = x^2 - 4x + 4$

8. $g(x) = 6 - 7x - x^2$

9. $h(x) = 2x^2 + 16x - 3$

10. $f(t) = -3t^2 + 6t + 5$

11. On page 150 we see that $f(x) = ax^2 + bx + c = a(x + \frac{b}{2a})^2 + c - \frac{b^2}{4a}$. The RHS here shows that the y-value of the vertex is given by $k = c - \frac{b^2}{4a}$. Evaluate $f(\frac{-b}{2a})$ to show that this is correct.

For numbers 12 - 14, show all work and print the answer in the space provided.

12. Which of the following is the standard form of the quadratic function $f(x) = 2x^2 + 16x + 15$?

A.) $f(x) = 2(x - 8)^2 - 113$ B.) $f(x) = 2(x + 4)^2 - 1$

C.) $f(x) = 2(x + 4)^2 - 17$ D.) $f(x) = 2(x + 8)^2 - 113$

E.) None of these.

Answer:_____

13. A quadratic function $h(x)$ has a leading coefficient of $a = -4$, has a y-intercept at $(0, 10)$ and the x-value of the vertex is 2. Which of the following represents $h(x)$?

A.) $h(x) = -4x^2 + 10x - 16$ B.) $h(x) = -4x^2 - 8x + 10$

C.) $h(x) = -4x^2 - 16x + 10$ D.) $h(x) = x^2 - 4x - 4$

E.) None of these.

Answer:_____

14. In which of the following are the vertex and x-value of the x-intercepts of $g(x) = -x^2 + 4x + 5$ correctly given?

A.) $V : (4, 5), x = 5$ B.) $V : (2, 13), x = 1, 5$ C.) $V : (-2, -7), x = -1, 5$

D.) $V : (2, 13), x = -1, 5$ E.) None of these.

Answer:_____

Chapter 17

Polynomial Functions

17.1   Notes about Polynomial Functions

Definition: A *polynomial function* is a function of the form
$P(x) = a_n x^n + a_{n-1} x^{n-1} + \ldots + a_1 x + a_0$, where $a_n \neq 0$, $n \geq 0$ and n is called
the *degree* of the function. $a_n, a_{n-1} \ldots, a_1$, and a_0 are called the *coefficients* of the
polynomial. a_n is the *leading coefficient*, $a_n x^n$ is the *leading term* and a_0 is the
constant coefficient or *constant term*.

Definition: The graph of any polynomial function is *smooth* curve (though smooth-
ness doesn't imply that the graph represents a polynomial). A graph that is continuous (has no breaks) and has no sharp points (kinks or corners) is smooth.

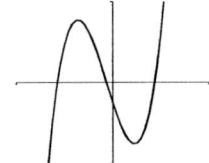

Figure 17.1: Smooth
graph.

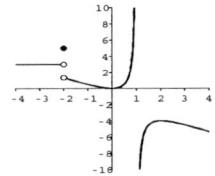

Figure 17.2: Not a
smooth graph.

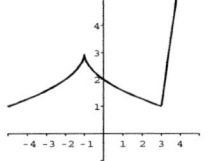

Figure 17.3: Not a
smooth graph.

Figure 17.1 shows the graph of a smooth function, while Figures 17.2 and 17.3
show graphs of functions that are not smooth. Figure 17.2 is not smooth because
it has breaks at $x = -2$ and 1. Figure 17.3 is not smooth because of the kink at
$x = -1$ and the corner at $x = 3$.

The graphs below will be used to illustrate general properties of polynomials.

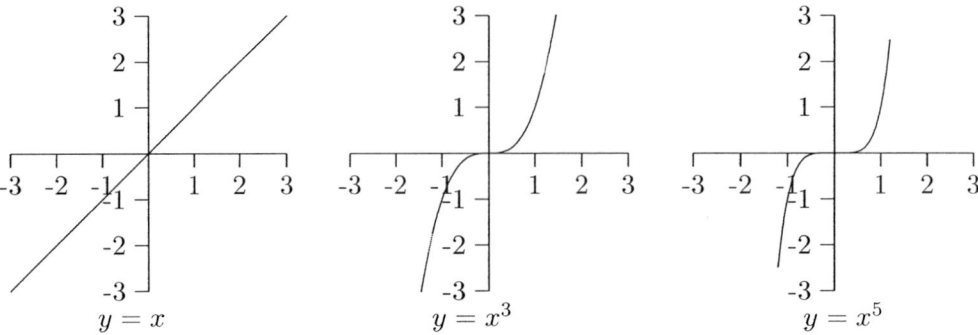

Note that each graph of a polynomial of odd degree is always increasing from left to right, haa a similar S-like shape (for degree > 1) and goes through the points $(-1, -1)$, $(0, 0)$, and $(1, 1)$.

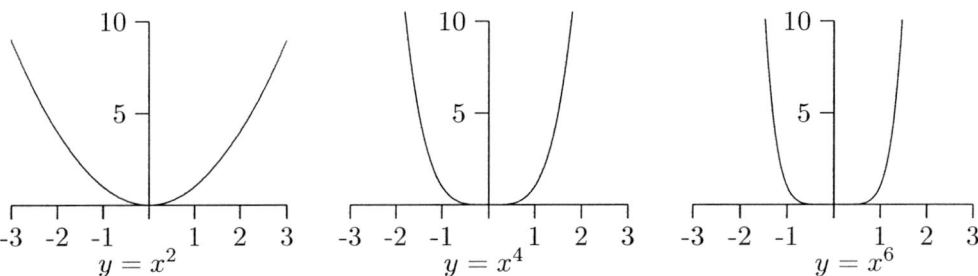

Each of the graphs of the polynomials of even degree decreases from $-\infty$ to 0 and increases from 0 to ∞, each has a similar U-like shape, and each also goes through the points $(-1, 1)$, $(0, 0)$, and $(1, 1)$.

Note the relative behavior of $y = x^3$ and $y = x^5$ and $y = x^2$ and $y = x^4$ between $(0, 0)$ and $(1, 1)$ and after $(1, 1)$ on the graphs below. The graphs of x^3 and x^2 are farther from the x-axis between 0 and 1 and closer afterwards. In general, the lower the degree of the monomial, the farther away the x-axis it will be between 0 and 1 and closer to the x-axis after 1. Similar behavior is true to the left of $x = 0$.

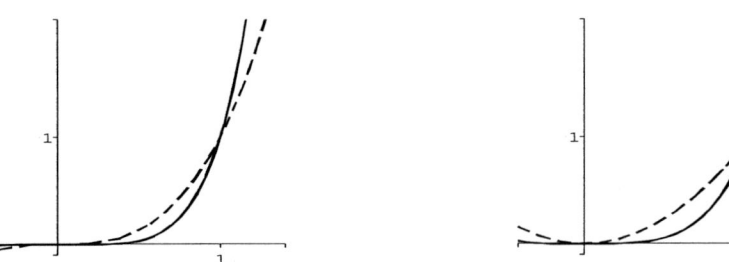

Figure 17.4: $y = x^3$ (dashed) and $y = x^5$ (solid).

Figure 17.5: $y = x^2$ (dashed) and $y = x^4$ (solid).

17.2 End behavior

Next we will examine the *end behavior* of polynomials. The end behavior refers to the part of the graphs after all twists and turns have occurred toward the 'center' of the graph.

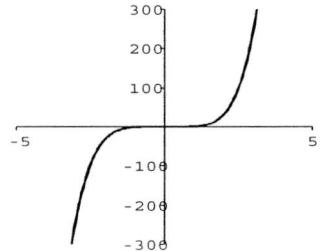

Figure 17.6: $y = P(x)$.

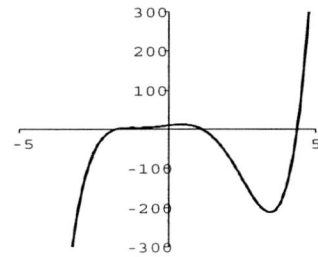

Figure 17.7: $y = Q(x)$.

Two views of the graphs of $P(x) = x^5$ and $Q(x) = x^5 - 5x^4 + 3x^4 - 10x^3 - 4x^2 + 9x + 10$ are shown. In the first view above, the graphs seem very different because of $y = Q(x)$ having several additional twists and turns. In the second view below, the graphs are indistinguishable. Because of this, we would say that the two graphs have the same end behavior.

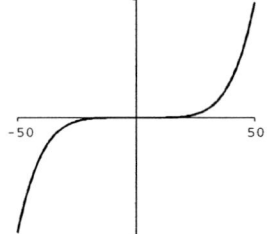

Figure 17.8: $y = P(x)$.

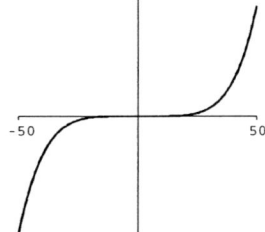

Figure 17.9: $y = Q(x)$.

End-behavior is determined entirely by the leading term because it 'swamps' the other terms as the x-values get very large or very small. The end behavior for both $y = P(x)$ and $y = Q(x)$ are due to the x^5-terms. The chart on the next page describes the 4 possible end behaviors and what the sign and degree of the leading terms would have to be to cause each end behavior.

End Behavior		
Polynomial with	As $x \to -\infty$	As $x \to \infty$
Even-degree & Positive Leading Coefficient	$y \to \infty$	$y \to \infty$
Even-degree & Negative Leading Coefficient	$y \to -\infty$	$y \to -\infty$
Odd-degree & Positive Leading Coefficient	$y \to -\infty$	$y \to \infty$
Odd-degree & Negative Leading Coefficient	$y \to \infty$	$y \to -\infty$

Figure 17.10: LC: Positive and Even.   Figure 17.11: LC: Negative and Even.

Figure 17.12: LC: Positive and Odd.   Figure 17.13: LC: Negative and Odd.

17.3   Zeros

Definition: The graph of the function $y = f(x)$ has a *zero* or *root* at $x = c$ if and only if $f(c) = 0$. A polynomial that has a zero at $x = c$ must have a *factor* of $(x-c)$.

Definition: The *multiplicity* of a zero refers to the power to which the factor which caused the zero is raised to in the polynomial.

For instance, $g(x) = (x - 5)^3(x + 4)^6$ has zeros at $x = 5$ and $x = -4$. The zero at $x = 5$ has multiplicity 3 and the zero at $x = -4$ has multiplicity 6. The whether the multiplicity of a zero is even or odd determines how it affects the graph of the function. If the zero is even, the graph 'bounces' off the zero and has a parabola-like shape there. If the zero is odd, the graph 'goes through' the zero and has an odd-powered shape there (either like a line or cubic function). See the graph of $g(x) = (x - 5)^3(x + 4)^6$ below.

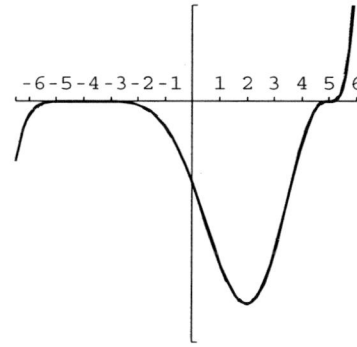

Figure 17.14: $y) = (x - 5)^3(x + 4)^6$.

17.4 ❖ Errors to Avoid

Note that the leading coefficient does not need to be the coefficient of the first term and the constant coefficient does not need to be the last term. In $g(x) = 3x^2 + 4 - x^5 + 6x$, the leading coefficient is -1 (because -1 is the coefficient of the term with the highest degree (x^5)) and the constant coefficient is 4. Also, not all polynomials have a constant coefficient. For example, $h(x) = 3x^2 + 5x^7 - 6x + 5x^3$ has no constant coefficient.

17.5   Examples

EX 1. Sketch the graph of $f(x) = -(x-1)(x+2)^2(x+5)^3$.

Solution ☞ To sketch this graph, we will determine the end behavior, y-intercept, zeros and the multiplicity of each zero. Because this polynomial isn't multiplied out, we first need to determine the sign and degree of the leading term to determine the end behavior. This can be done by determining the term of highest degree of each product were it already raised to the appropriate power and then multiplying these out. The appropriate terms from each product are -1, x, x^2 and x^3, respectively, meaning $-1 \cdot x \cdot x^2 \cdot x^3 = -x^6$ is the leading term of the polynomial. Thus as $x \to -\infty$, $y \to -\infty$ and as $x \to \infty$, $y \to -\infty$.

To find the y-intercept, it is sufficient to find $f(0)$ which is $f(0) = -(0-1)(0+2)^2(0+5)^3 = 500$. So the y-intercept is at $(0, 500)$.

To find the zeros we need only set each product equal to zero giving $x - 1 = 0$, $(x+2)^2 = 0$ $(x+5)^3 = 0$. This yields that 1, -2 and -5 are the zeros with multiplicity 1, 2 and 3, respectively.

The graph below (at left) shows the graph with the appropriate points and end behavior.

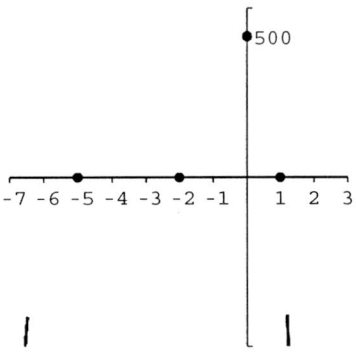

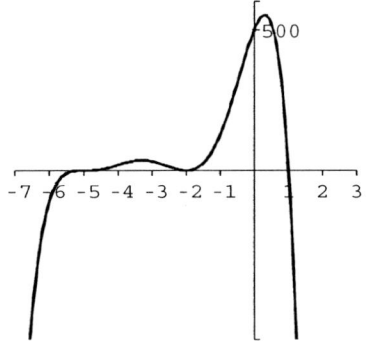

Figure 17.15: Some points on $y = -(x-1)(x+2)^2(x+5)^3$.

Figure 17.16: $y = -(x-1)(x+2)^2(x+5)^3$.

Further, we know that the graph goes through the points $(1, 0)$ and $(-5, 0)$ and bounces off at $(-2, 0)$ and connects these points as well as 'connecting' to the end behavior. This yields the graph above (at right).

EX 2. A polynomial function f has zeros only at $x = -2, 3$ and 1. The function has the following end behavior: as $x \to -\infty$, $y \to -\infty$ and as $x \to \infty$, $y \to -\infty$. If the zero at $x = 3$ has multiplicity 2 and the zeros at $x = -2$ and 1 have multiplicity 1 and a y-intercept at $(0, 4)$, find $f(x)$.

Solution ☞ Because the function has zeros at $x = -2, 3$ and 1, the polynomial must have factors $x+2$, $x-1$ and $x-3$. Because $x = 3$ has multiplicity 2, the $x - 3$ factor must be squared. Thus, $f(x) = a(x + 2)(x - 1)(x - 3)^2$, for some value of the constant a. Because the y-intercept is known to be at $(0, 4)$, we can find a by substituting $x = 0$ and $f(0) = 4$ into the function yielding $4 = a(0 + 2)(0 - 1)(0 - 3)^2 \Rightarrow 4 = 18a \Rightarrow a = 4/18 = 2/9$. Thus, $f(x) = (2/9)(x + 2)(x - 1)(x - 3)^2$. The graph below of $y = (2/9)(x + 2)(x - 1)(x - 3)^2$ shows that the graph has the correct zeros and end behavior.

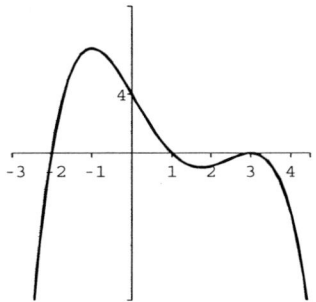

Figure 17.17: $y = f(x)$.

EX 3. ✭ (S01#14) Match the polynomial function $P(x) = -x^2 \left(x^2 - 4\right)$ with one of the following graphs.

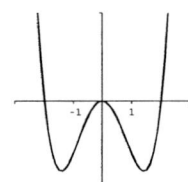

A.

B.

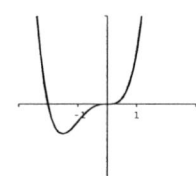

C.

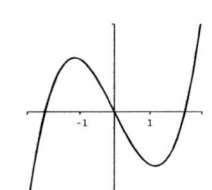

D.

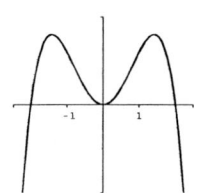

E.

Solution ☞ The graph of $y = P(x)$ should have zeros at $x = 0, -2$ and 2, with multiplicity 2, 1, and 1, respectively. The end behavior of the polynomial should be as $x \to -\infty$, $y \to -\infty$ and $x \to \infty$, $y \to -\infty$. The only graph meeting all of these criteria is E.).

Chapter 17 Exercises

Sketch the graphs of the functions described in 1 - 4 on the graphs provided. Label all zeros.

1. The leading term is $-x^4$ and the graph's only zeros are $x = -5, -2, 1$, and 4.

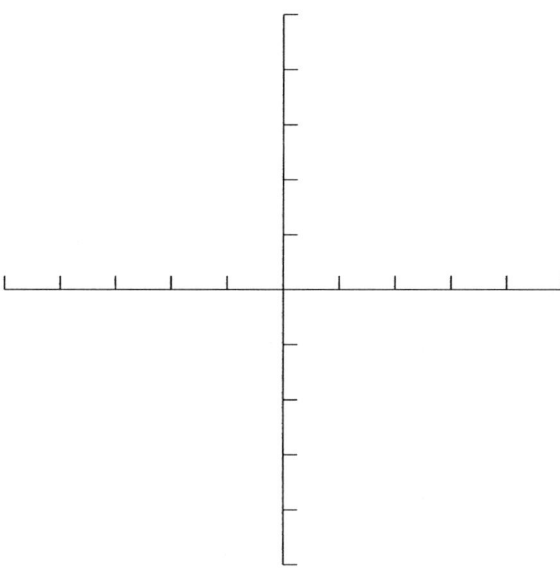

2. The leading term is $-x^7$ and the graph's only zeros are $x = -4, -1$, and 2 and the multiplicity of the zero at $x = 2$ is 5.

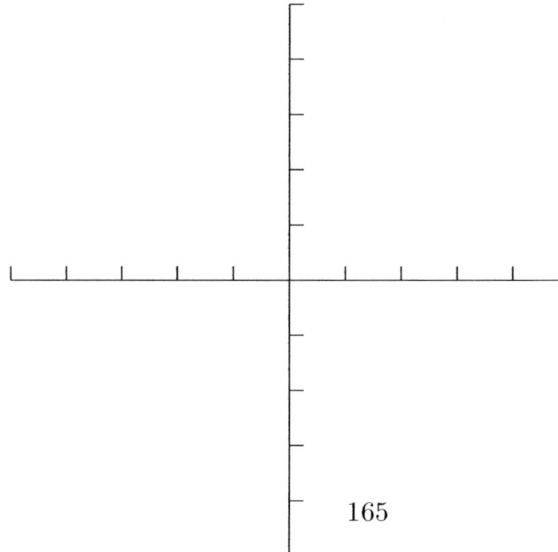

165

3. Has factors $x - 3$, $x + 2$ and $x - 4$ with multiplicities 2, 4, and 3, respectively. The end behavior is as $x \to -\infty$, $y \to \infty$ and as $x \to \infty$, $y \to -\infty$.

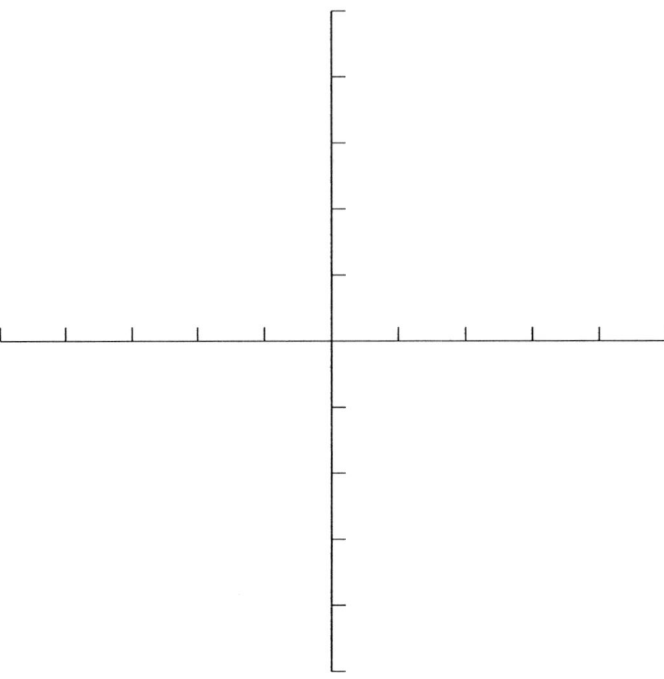

4. Has a y-intercept at $(0, -2)$, zeros at $x = -2, 1$, and 4 and the zero at 1 has degree 2. The end behavior is as $x \to -\infty$, $y \to \infty$ and as $x \to \infty$, $y \to \infty$.

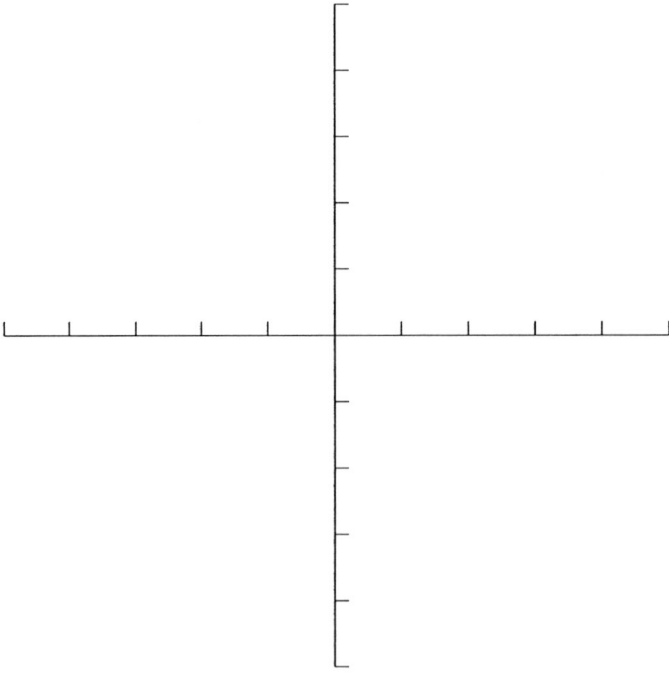

Determine a function that satisfies the conditions described in each of 5 - 8. More than one answer is possible on many of these problems.

5. The leading term is $-x^4$ and the graph's only zeros are $x = -5, -2, 1$, and 4

6. The leading term is $-3x^7$ and the graph's only zeros are $x = -4, -1$, and 2 and the multiplicity of the zero at $x = 2$ is 5.

7. Has a y-intercept at $(0, -4)$, zeros at $x = -4, 2$, and 3 and the zeros at $x = 2$ and 3 each have degree 1. The end behavior is as $x \to -\infty$, $y \to -\infty$ and as $x \to \infty$, $y \to -\infty$.

8. Zeros at $x = -5, -2$, and 1 and the zero at -2 has multiplicity 2. y-intercept at $(0, 10)$ and end behavior $x \to -\infty$, $y \to -\infty$ and as $x \to \infty$, $y \to \infty$.

Use the graph below to answer questions 9 - 11.

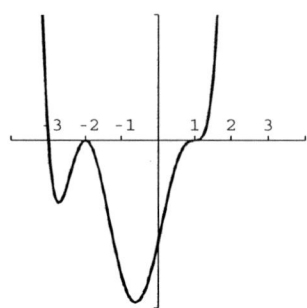

Figure 17.18: $y = g(x)$

9. Which of the zeros in $y = g(x)$ have odd multiplicity? Which have even multiplicity?

10. Is the leading coefficient of $y = g(x)$ positive or negative? Is the degree of the polynomial even or odd?

11. Describe the end behavior of $y = g(x)$ in the space below.

   As $x \to -\infty$, $y \to$ _____.

   As $x \to \infty$, $y \to$ _____.

For numbers 12 - 14, show all work and print the answer in the space provided.

12. Determine the end behavior of
   $P(x) = 5(x^3 - 5x^2 + 7x)^5(x^2 + 2)^4(x^7 - 6x^4 + 5x^2)$.

   A.) As $x \to -\infty, y \to -\infty$
       As $x \to \infty, y \to -\infty$

   B.) As $x \to -\infty, y \to -\infty$
       As $x \to \infty, y \to \infty$

   C.) As $x \to -\infty, y \to \infty$
       As $x \to \infty, y \to -\infty$

   D.) As $x \to -\infty, y \to \infty$
       As $x \to \infty, y \to \infty$

   E.) None of the above.

   Answer:_____

13. The zeros (Z) and multiplicity (M) of $g(x) = -6(x^2 + 6x + 9)(x - 2)(x^4)$ are given in which of the following?

   A.) $Z : -3, 2, 0; M : 2, 1, 4$

   B.) $Z : 6, 2, 0; M : 1, 1, 1$

   C.) $Z : -3, 2, 1; M : 2, 1, 4$

   D.) $Z : 3, 2, 0; M : 2, 1, 4$

   E.) None of these.

   Answer:_____

14. The graph below is the graph of which of the following functions?

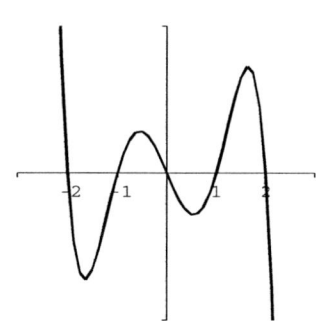

   A.) $f(x) = x(x^2 - 1)(x^2 - 4)$

   B.) $f(x) = -x^2(x^2 - 1)(x^2 - 4)$

   C.) $f(x) = -x(x^2 - 4)(x^2 - 1)$

   D.) $f(x) = -x(x - 1)^2(x - 2)^2$

   E.) None of these.

   Answer:_____

Chapter 18

Division of Polynomials

18.1 Notes about Division of Polynomials

Definition: The *division algorithm* states that for any two polynomials $P(x)$ and $D(x)$ with degrees m and n, respectively, with $m \geq n$, there exist unique polynomials $Q(x)$ and $R(x)$, so that $P(x) - D(x) \cdot Q(x) + R(x)$, with the degree of $R(x) < n$ or $R(x) = 0$. If $R(x) = 0$, then $D(x)$ is a factor of $P(x)$. $P(x)$ is called the *dividend*, $D(x)$ is called the *divisor*, $Q(x)$ is called the *quotient*, and $R(x)$ is called the *remainder*.

Long division of polynomials works in a similar way to long division of integers. See the **Examples** section for examples showing how it works.

Synthetic division is a method of polynomial division to be used when the dividend is of the form $x - c$, where c is a real number. Otherwise, long division must be used. This method involves using an abstraction of the given polynomial as shown below. If $f(x) = ax^n + bx^{n-1} + \ldots + d$ is divided by $x - c$, the method of solution using synthetic division is shown below.

$$
\begin{array}{c|ccccc}
c & a & b & \ldots & \ldots & d \\
  &   & ac & \ldots & \ldots & cz \\
\hline
  & a & b+ac & \ldots & z & d+cz
\end{array}
$$

The values inside the 'box' on the top line represent the coefficients of the dividend polynomial. All but the last of the values below the horizontal line represent the coefficients of the quotient polynomial, while the last value represents the remainder. The remainder of a synthetic division problem is always a real number.

The work above shows that $\dfrac{f(x)}{x - c} = ax^{n-1} + (b + ac)x^{n-2} + \ldots + z + \dfrac{d + cz}{x - c}$.

The Remainder Theorem: If $P(x)$ is divided by $x - k$, then the remainder is given by $P(k)$.

18.2 ❖ Errors to Avoid

Be careful when using long division to *subtract*, rather than add, the terms that had been multiplied by the divisor.

Be careful when using *synthetic* division to *add*, rather than subtract, the terms that had been multiplied by c.

Note that the c in $x - c$ can be negative. See Example 3.

18.3 Examples

EX 1. Divide $f(x) = 5x^3 + 4x^2 - 10$ by $x^2 + x - 1$.

Solution ☞ This division requires long division because the divisor is not of the form $x - c$.

$$
\begin{array}{r}
5x \\
x^2 + x - 1 \overline{)\ 5x^3 + 4x^2 + 0x - 10}
\end{array}
$$

First, write the dividend under the long division symbol and the divisor to the left. To determine the first term of the quotient, divide the first term of the dividend by the first term of the divisor which gives $\dfrac{5x^3}{x^2} = 5x$. Write this above the horizontal line.

$$
\begin{array}{r}
5x \\
x^2 + x - 1 \overline{)\ 5x^3 + 4x^2 + 0x - 10} \\
-(5x^3 + 5x^2 - 5x) \\
\hline
-x^2 + 5x - 10
\end{array}
$$

Next multiply the $5x$ times the divisor. This yields $5x^3 + 5x^2 - 5x$ and should be written below the dividend. Subtract the result from the dividend and write this below the $5x^3 + 5x^2 - 5x$-term.

$$
\begin{array}{r}
5x - 1 \\
x^2 + x - 1 \overline{)\ 5x^3 + 4x^2 + 0x - 10} \\
-(5x^3 + 5x^2 - 5x) \\
\hline
-x^2 + 5x - 10 \\
-(-x^2 - x + 1) \\
\hline
6x - 11
\end{array}
$$

These steps will be repeated until the result of the subtraction has a degree less than the divisor. In this case, this is only one more step. $\dfrac{-x^2}{x^2} = -1$, so this is put above the horizontal line and then multiplied by the divisor and subtracted from the result, leaving $6x - 11$.

This division shows that $5x^3 + 4x^2 - 10 = (x^2 + x - 1)(5x - 1) + (6x - 11)$.

EX 2. ✭ (F02#7) Find the quotient (q) and remainder (r) if $x^5 + x^4 + x^2 + 1$ is divided by $x^2 + x + 1$.

A.) $q = x^3; r = x - 1$ B.) $q = x^3 - x + 2; r = -x - 1$

C.) $q = x^3; r = 1$ D.) $q = x^3 + 2x^2 + 3x + 6; r = 9x + 7$

E.) $q = x^2 + 2x + 3; r = 4$

Solution ☞ For this problem, long division is again necessary because the divisor is not of the form $x - c$. The long division is shown below.

$$
\begin{array}{r}
x^3 \qquad - \quad x + 2 \\
x^2 + x + 1 \overline{)\quad x^5 + x^4 + 0x^3 + \quad x^2 + 0x + 1} \\
-\underline{(x^5 + x^4 + \quad x^3)} \\
- \quad x^3 + \quad x^2 + 0x + 1 \\
-\underline{(- \quad x^3 - \quad x^2 - \quad x\,)} \\
2x^2 + \quad x + 1 \\
-\underline{(\quad 2x^2 + 2x + 2\,)} \\
- \quad x - 1
\end{array}
$$

Thus, the quotient is $x^3 - x + 2$ and the remainder is $-x - 1$ and the answer is B.).

EX 3. Use synthetic division to divide $g(x) = 2x^4 + 5x^3 - 6x + 12$ by $x + 4$.

Solution ☞

$$
\begin{array}{r|rrrrr}
-4 & 2 & 5 & 0 & -6 & 12 \\
   &   &   &   &   &   \\
\hline
   & 2 &   &   &   &   \\
\end{array}
$$

For the initial setup, the coefficients of the dividend are along the top line of the box. Bring the left most value down below the horizontal line.

$$
\begin{array}{r|rrrrr}
-4 & 2 & 5 & 0 & -6 & 12 \\
   &   & -8 & &   &   \\
\hline
   & 2 & -3 & &   &   \\
\end{array}
$$

Next multiply -4 by 2. Place this value above the horizontal line under the next coefficient on the top line, 5 in this case. Then add -8 to 5, yielding -3. Place this result below the horizontal line.

$$
\begin{array}{r|rrrrr}
-4 & 2 & 5 & 0 & -6 & 12 \\
   &   & -8 & 12 & -48 & 216 \\
\hline
   & 2 & -3 & 12 & -54 & 228 \\
\end{array}
$$

Multiply -4 by -3 and place the 12 under the 0 and then take 0 plus 12 and place the result below the horizontal line. Continue multiplying and adding until none of the coefficients of the dividend are left.

The numbers under the horizontal line represent the coefficients of the quotient and the remainder. The resulting quotient is $2x^3 - 3x^2 + 12x - 54$ and the remainder is 228. Put another way $2x^4 + 5x^3 - 6x + 12 = (x + 4)(2x^3 - 3x^2 + 12x - 54) + 228$.

✎ Though it is not a guarantee that the answer is correct, using the remainder theorem can help make you more confident that the answer is correct. Finding $g(-4) = 2(-4)^4 + 5(-4)^3 - 6(-4) + 12 = 512 - 320 + 24 + 12 = 228$.

EX 4. ★ (S04#10) Find the constant c so that the denominator of $\dfrac{x^3 - 4x^2 + 7x + c}{x - 3}$ will divide evenly into the numerator?

A.) -12 B.) 3   C.) 84   D.) 7   E.) -30

Solution ☞ To do this, we will use synthetic division to determine the remainder as a function of c. The appropriate division is shown below.

$$
\begin{array}{r|rrrr}
3 & 1 & -4 & 7 & c \\
  &   & 3 & -3 & 12 \\
\hline
  & 1 & -1 & 4 & c + 12 \\
\end{array}
$$

We see that the remainder is $c + 12$. If $x - 3$ divides evenly into $x^3 - 4x^2 + 7x + c$, then the remainder must be 0. Thus, $c + 12 = 0 \Rightarrow c = -12$ or A.).

EX 5. Use $f(x) = x^3 + 5x^2 - 2x - 24$ to answer the questions that follow.

(a) Show that $x = 2$ is a zero of $f(x)$ using the Remainder Theorem.

Solution ☞ If $x = 2$ is a zero, then the remainder will be 0. To make use of the Remainder Theorem, we must find $f(2)$ which yields $f(2) = (2)^3 + 5(2)^2 - 2(2) - 24 = 8 + 20 - 4 - 24 = 0$. Thus, $x = 2$ is a zero of $f(x)$.

(b) Completely factor $f(x)$.

Solution ☞ If $x = 2$ is a zero of $f(x)$, then $x - 2$ is a factor of $f(x)$. Thus, synthetic division can be used to further factor $f(x)$. The results of the division are shown below.

$$
\begin{array}{r|rrrr}
2 & 1 & 5 & -2 & -24 \\
  &   & 2 & 14 & 24 \\
\hline
  & 1 & 7 & 12 & 0
\end{array}
$$

We see that $f(x) = (x-2)(x^2+7x+12)+0$. Finishing the factoring yields
$$f(x) = (x - 2)(x + 4)(x + 3).$$

Chapter 18 Exercises

Use Long Division to perform the indicated division for problems 1 - 4. Clearly indicate the quotient and remainder in your answer.

1. Divide $f(x) = x^3 - 4x^2 + 5x - 7$ by $x^2 + 3x - 4$.

2. Divide $g(x) = 3x^4 - 4x^2 - 12x + 1$ by $x^2 - 5$

3. Divide $h(x) = 5x^4 - 4x^3 + 3x^2 - 2x + 1$ by $x^3 + 5x^2 - 3$

4. Divide $t(x) = x^7 + x^5 + x^3 + x - 1$ by $x^2 + x$

Use Synthetic Division to perform the indicated division for problems 5 - 8. Clearly indicate the quotient and remainder in your answer.

5. Divide $f(x) = x^3 - 4x^2 + 8x - 10$ by $x - 2$

6. Divide $s(t) = 5t^4 + 5t^3 - 4t^2 + 4t + 19$ by $t + 1$

7. Divide $g(x) = x^5 - 2x^3 - 15x^2 + 5$ by $x - 3$

8. Divide $h(x) = x^5 - 625$ by $x + 5$

Use Synthetic Division to determine which of the given values are zeros of the given polynomial and then completely factor the polynomial over the complex numbers for problems 9 - 13.

9. $f(x) = x^2 - 4x + 5$; $x = 1, 2, 3$

10. $g(x) = x^3 - 6x^2 - 7x + 60; \; x = -2, 2, 4$

11. $h(x) = x^4 - 5x^3 - 25x^2 + 65x + 84; \; x = -1, 3, 4$

12. $k(x) = x^5 - 2x^4 + x^3 - 2x^2; \; x = -2, 2, 4$

13. $l(x) = x^4 - x^3 - 17x^2 + 55x - 50$; $x = 1, 2, 5$

14. Use the Remainder Theorem to determine if $x - 1$ is a factor of $R(x) = x^{900} - 765x^{732} + 770x^{523} + 6$. Show all work.

For numbers 15 - 18, show all work and print the answer in the space provided.

15. What is the remainder when $x^5 - 3x^4 + 5x^2 + 10$ is divided by $x^2 - 3x + 1$?

    A.) $7x + 8$ B.) $x^3 - x + 2$ C.) $12x + 6$

    D.) 0 E.) None of these.

    Answer:_____

16. A correctly performed synthetic division is shown below.

$$
3 \begin{array}{|rrrr} 1 & -7 & 10 & 11 \\ & 3 & -12 & -6 \\ \hline 1 & -4 & -2 & 5 \end{array}
$$

Which of the following represent the quotient (Q) and remainder (R)?

A.) $Q = x^2 - 4x - 2; R = 5$ B.) $Q = x^3 - 7x^2 + 10x + 11; R = x - 3$

C.) $Q = x - 3; R = x - 5$ D.) $Q = 3x^2 - 12x - 6; R = x^3 - 4x^2 - 2x + 5$

E.) None of these.

Answer:_____

17. Find the constant c so that $x + 1$ will divide evenly into $5x^3 + 8x^2 + cx - 2$.

A.) 0 B.) -1 C.) 5 D.) 1 E.) None of these.

Answer:_____

18. Find a constant c so that $x - c$ will divide evenly into $x^3 - 216$.

    A.) -6 B.) 216   C.) 36
    D.) 6   E.) None of these.

                                        Answer:_____

19. What is the remainder of $P(x) = x^{10} - 1032$ when divided by $x - 2$?

    A.) 1030   B.) -8 C.) 1056
    D.) 8   E.) None of these.

                                        Answer:_____

Chapter 19

Finding Zeros of Polynomials

19.1 Notes about Finding Zeros of Polynomials

In this chapter, several theorems are given. These theorems relate to finding zeros of polynomials or factoring polynomials.

19.1.1 Limiting Theorems

The Rational zeros theorem, Descartes' rule of signs, and the Upper and Lower bounds theorem all help narrow down what the zeros could be, thereby limiting what must be checked do help find zeros.

Rational zeros theorem: If the polynomial
$P(x) = a_n x^n + a_{n-1} x^{n-1} + \ldots + a_1 x + a_0$ has integer coefficients, then every rational zero of P is of the form $\dfrac{t}{b}$, where t is a factor of the constant coefficient a_0 and b is a factor of the leading coefficient a_n.

The rational zeros theorem limits what rational values must be searched to find the zeros and often helps by giving a starting point for what values to check (see Examples 1, 7 and 2). No other rational zeros are possible.

Descartes' rule of signs: Let P be a polynomial with real coefficients.

1. The number of positive real zeros of $P(x)$ is either equal to the number of variations in sign in $P(x)$ or is less than that by an even integer.

2. The number of negative real zeros of $P(x)$ is either equal to the number of variations in sign in $P(-x)$ or less than that by an even integer.

Descartes' rule of signs has various uses but a primary use of telling us that we have found all of the zeros (see Example 4a).

Upper and lower bounds theorem: Let P be a polynomial with real coefficients.

1. If we divide $P(x)$ by $x - b$ (with $b > 0$) using synthetic division, and if the entries in the row that contains the quotient and remainder has no negative entry, then b is an upper bound for the real zeros of P.

2. If we divide $P(x)$ by $x + a$ (with $a > 0$) using synthetic division, and if the entries in the row that contains the quotient and remainder alternate between nonpositive and nonnegative, then a is a lower bound for the real zeros of P.

The upper and lower bounds theorem helps narrow down the portion of the real number line that needs to be checked to find all of the real zeros (see Example 5 and 10).

19.1.2   Existence Theorems

The following theorems all fall into the category of existence theorems, or theorems that describe the circumstances that guarantee that an object exists, but do not describe how to find the object.

Fundamental theorem of algebra: Every polynomial of degree $n \geq 1$ $P(x) = a_n x^n + a_{n-1} x^{n-1} + \ldots + a_1 x + a_0$ with complex coefficients has at least one complex zero.

The fundamental theorem of algebra forms the basis for the proof of the remaining theorems and has significant historical importance.

Complete factorization theorem: If $P(x)$ is a polynomial of degree $n \geq 1$, then there exist complex numbers $a, c_1, c_2, \ldots, c_{n-1}, c_n$ (with $a \neq 0$) such that $P(x) = a(x - c_1) \cdot (x - c_2) \cdots (x - c_{n-1}) \cdot (x - c_n)$.

The complete factorization theorem describes an important phenomenon that is extremely useful in factoring polynomials. In particular, it tells us the necessary conditions to be done factoring.

Zeros theorem: Every polynomial of degree $n \geq 1$ has exactly n complex zeros, provided that a zero of multiplicity k is counted k times.

The zeros theorem succinctly states a consequence of the complete factorization theorem.

Conjugate zeros theorem: If the polynomial P has real coefficients, and if the complex number z (with imaginary part not equal to zero) is a zero of P, then its complex conjugate, $\overline{z}$, is also a zero of P.

The conjugate zeros theorem adds significant information when the polynomial has real coefficients and cannot be completely factored over the reals.

Linear and quadratic factors theorem: Every polynomial with real coefficients can be factored into a product of linear and irreducible quadratic factors with real coefficients.

Along with the complete factorization theorem, describes the sort of situations that can come about while factoring.

19.2 ❖ Errors to Avoid

Be careful to note whether the theorems deal with complex numbers or real numbers.

Remember that all real numbers are complex numbers, but not all complex numbers are real.

When using the upper and lower bounds theorem, remember that 0 can be considered nonpositive or nonnegative, depending on the situation.

Not all polynomials have constant coefficients. To see how to use the rational zeros theorem on a polynomial without a constant coefficient, see Example 2.

19.3 Examples

EX 1. List all possible rational zeros of $R(x) = 2x^5 + 3x^3 + 4x^2 - 8$ according to the Rational Zeros Theorem.

Solution ☞ The integer factors of -8 are $\pm 1, \pm 2, \pm 4$, and ± 8 and the integer factors of 2 are ± 1 and ± 2. Dividing each of the factors of -8 by each of the factors of 2 yield $\pm 1, \pm 2, \pm 4, \pm 8$, and $\pm 1/2$.

EX 2. Find all possible rational zeros of $P(x) = 7x^7 - 21x^5 - 35x^6 + 105x^4 - 28x^3 + 140x^2$.

Solution ☞ First note that $P(x) = x^2 \cdot Q(x)$ for $Q(x) = 7x^5 - 21x^4 - 35x^3 + 105x^2 - 28x + 140$. Now the rational zeros theorem can be applied to $Q(x)$ to yield $\pm 1, \pm 2, \pm 5, \pm 7, \pm 10, \pm 14, \pm 20, \pm 28, \pm 70, \pm 140, \pm 1/7, \pm 2/7, \pm 5/7, \pm 10/7, \pm 20/7$, and ± 4.

EX 3. Determine the possible number of positive and negative real zeros of $f(x) = x^4 + 5x^3 + 7x^2 + 5x + 1$ according to Descartes' rule of signs.

Solution ☞ The signs of the coefficients of $f(x)$ are $+, +, +, +$, and $+$. Thus, there are no changes of sign and no positive real zeros.

$f(-x) = x^4 - 5x^3 + 7x^2 - 5x + 1$ and the signs of its coefficients are $+, -, +, -$, and $+$ and there are 4 changes in sign. Thus, there are either 4, 2 or 0 negative real zeros. In this case, there are 2 negative real zeros, $\approx -3.32, -.30$.

EX 4. Use $P(x) = 2x^6 + 5x^4 - x^3 - 5x - 1$ to answer the questions that follow.

(a) Use Descartes' Rule of Signs to determine how many positive and how many negative real zeros $P(x)$ can have.

Solution ☞ The signs of the coefficients of $P(x)$ in order are $+, +, -, -$, and $-$ and there is only 1 change in sign (which is underlined). Therefore, there is only 1 positive real root.

$P(-x) = 2x^6 + 5x^4 + x^3 + 5x - 1$ and thus the signs of the coefficients of $P(-x)$ in order are $+, +, +, +$, and $-$ and there is only 1 change in sign. Therefore, there is only 1 negative real root.

(b) Find all rational zeros of $P(x)$.

Solution ☞ Noting that the possible rational zeros are ± 1 and $\pm 1/2$, we can begin to check these using the remainder theorem. $P(1) = 2(1)^6 + 5(1)^4 - (1)^3 - 5(1) - 1 = 2 + 5 - 1 - 5 - 1 = 0$. Thus, $x = 1$ is a zero and we know by Descartes' rule of signs that it is the only positive real zero.

Checking the negative values, we see that $P(-1) = 2(-1)^6 + 5(-1)^4 - (-1)^3 - 5(-1) - 1 = 2 + 5 + 1 + 5 - 1 \neq 0$ and $P(-1/2) = 2(-1/2)^6 + 5(-1/2)^4 - (-1/2)^3 - 5(-1/2) - 1 = 0.031250 + .3125 + 0.125 + 2.5 - 1 \neq 0$. Thus, $x = 1$ is the only rational zero of $P(x)$.

EX 5. Find integers that are upper and lower bounds for the real zeros of the polynomial: $P(x) = 2x^3 - 3x^2 - 8x - 12$.

Solution ☞ Before using the synthetic division necessary to make use of the upper and lower bounds theorem, we'll use the remainder theorem to find an appropriate positive starting value. Noting that $P(1) = -21$, $P(2) = -24$, $P(3) = -9$, and $P(4) = 36$, the lowest positive value that could be a positive upper bound is 4 (because the remainder must be positive for all of the values to be positive. Using $b = 4$, synthetic division yields the following bottom row: $2, 5, 12, 36$. Because all of the values are positive, 4 is an upper bound for the real zeros.

Using $a = -1$, synthetic division yields the following bottom row: $2, -5, -3, -9$. Because the values do not alternate between positive and negative, the upper and lower bounds theorem doesn't tell us that -1 is a lower bound. Continuing with $a = -2$, synthetic division yields the bottom row: $2, -8, 8, -28$. Because the values alternate between positive and negative, the upper and lower bounds theorem tells us that -2 is a lower bound.

Thus, we know that 4 and -2 are upper and lower bounds of the real zeros of $P(x)$.

EX 6. Find a polynomial with integer coefficients that satisfy: $Q(x)$ has degree 3, and zeros -5 and $2 + 4i$.

Solution ☞ Knowing that $Q(x)$ has zeros at -5 and $2 + 4i$ means that $Q(x)$ has factors $x - (^-5)$ and $x - (2 + 4i)$. Further, the conjugate zeros theorem tells us that $2 - 4i$ is also a zero meaning $x - (2 - 4i)$ is a factor. Thus, $Q(x) = a(x -^- 5)(x - (2 + 4i))(x - (2 - 4i))$ for some constant a. Assuming $a = 1$, then $Q(x) = x^3 + x^2 + 100$.

EX 7. Find all real and complex zeros of the polynomial: $P(x) = x^5 + 3x^4 + 8x^3 + 24x^2 - 9x - 27$.

Solution ☞ The possible rational zeros are $\pm 1, \pm 3, \pm 9$, and ± 27. Using this as a starting point for what zeros to look for we see that $P(1) = 0$, so $x - 1$ is a factor. Using synthetic division, we see that $P(x) = (x-1)(x^4 + 4x^3 + 12x^2 + 36x + 27)$. For simplicity, we'll call $Q(x) = x^4 + 4x^3 + 12x^2 + 36x + 27$. Noting that there are no sign changes in $Q(x)$ means there are no positive real zeros. Checking negative values of the possible rational zeros (which are the same as $P(x)$'s), we see that $Q(-1) = 0$. Again using synthetic division, we see that $P(x) = (x-1)(x+1)(x^3 + 3x^2 + 9x + 27)$ which factors to $P(x) = (x-1)(x+1)(x^2(x+3) + 9(x+3)) = (x-1)(x+1)(x+3)(x^2 + 9)$. Solving $x^2 + 9 = 0$ yields $x = \pm 3i$. Thus, the zeros of $P(x)$ are $x = 1, -1, -3$, and $\pm 3i$.

EX 8. For the polynomial $P(x) = x^4 + 8x^2 - 9$,

(a) Factor P into linear and irreducible quadratic factors with real coefficients.

Solution ☞ If we let $u = x^2$ then $u^2 = x^4$ which means $P(u) = u^2 + 8u - 9 = (u - 9)(u + 1)$. Thus, $P(x) = (x^2 - 9)(x^2 + 1) = (x + 3)(x - 3)(x^2 + 1)$.

(b) Factor P completely into linear factors with complex coefficients.

Solution ☞ This only requires factoring $x^2 + 1$. Noting that this is the difference of two complex squares, we see that $x^2 + 1 = (x - i)(x + i) \Rightarrow P(x) = (x + 3)(x - 3)(x - i)(x + i)$.

EX 9. ★ (S04#21) Use $f(x) = 2x^3 - 2x^2 - 74x - 70$ to answer the questions that follow.

(a) According to the Rational Zeros Test, what are the possible rational zeros? List them all in the space provided.

Solution ☞ $\pm 1, \pm 2, \pm 5, \pm 7, \pm 10, \pm 14, \pm 35, \pm 70, \pm 1/2, \pm 5/2, \pm 7/2$, and $\pm 35/2$.

(b) Use the Remainder Theorem to show that $x = -1$ is a root of $f(x)$.

Solution ☞ $f(-1) = 2(-1)^3 - 2(-1)^2 - 74(-1) - 70 = -2 - 2 + 74 - 70 = 0$.

(c) Completely factor $f(x)$.

Solution ☞ Noting that $x = -1$ is a root, we'll use synthetic division to further factor $f(x)$. The bottom row of the synthetic division by $x = -1$ is $2, -4, -70$, and 0 meaning $f(x) = (x + 1)(2x^2 - 4x - 70)$. Factoring further yields $f(x) = 2(x + 1)(x^2 - 2x - 35) = 2(x + 1)(x - 7)(x + 5)$.

EX 10. ★ (F02#21) For the following questions, use $f(x) = 5x^3 - 15x^2 - 20x + 60$.

(a) Is $x = 3$ an upper bound for the zeroes of f according to the upper and lower bounds theorem? If not, find the smallest integer that is an upper bound according to the upper and lower bounds theorem.

Solution ☞ To determine if $x = 3$ is an upper bound, we need to perform synthetic division. Doing so yields the following bottom row: $5, 0, -20, 0$. Because the values are not all nonnegative, the theorem does not tell us that $x = 3$ is an upper bound. Performing synthetic division with $x = 4$ yields a bottom row of $5, 5, 0, 60$ that is all nonnegative and by the theorem, $x = 4$ is a upper bound. Moreover, because $x = 3$ isn't known to be an upper bound, $x = 4$ is the lowest value the theorem will tell us is an upper bound.

(b) Is $x = -1$ a lower bound for the zeroes of f according to the upper and lower bounds theorem? If not, find the largest integer that is a lower bound according to the upper and lower bounds theorem.

Solution ☞ To determine if $x = -1$ is an lower bound, we need to perform synthetic division. Doing so yields the following bottom row: $5, -20, 0, 60$. Because the values are not alternately nonpositive and nonnegative, the theorem does not tell us that $x = -1$ is a lower bound. Performing synthetic division with $x = -2$ yields a bottom row of $5, -25, 30, 0$ and we see that $x = -2$ is the greatest lower bound.

(c) Completely factor $f(x)$.

Solution ☞ $P(3) = 0$ which means that 3 is a zero of $P(x)$ and $x - 3$ is a factor. Further, based on the results of (10a) we see that $P(x) = (x - 3)(5x^2 - 20)$. Factoring this yields $P(x) = 5(x - 3)(x^2 - 4) = 5(x - 3)(x + 2)(x - 2)$.

Chapter 19 Exercises

Determine all of the possible rational zeros for 1 - 3 according to the rational zeros theorem.

1. $f(x) = x^3 - 6x + 10$

2. $g(x) = 22x^9 + 5x^6 - 3x^4 + 8x - 14$

3. $p(x) = 6x^4 - x^2 + 7x + 15$

For numbers 4-6 completely factor the given function into linear terms. Show all work.

4. $f(x) = 4x^4 + 3x^3 - 12x^2 - 17x - 6$

5. $g(x) = 6x^4 - 7x^3 - 26x^2 + 7x + 20$

6. $f(x) = -3x^4 - 9x^3 + 87x^2 + 225x - 300$

7. Use $f(x) = -x^4 + 4x^3 + x^2 + 16x + 20$ to answer the questions below.

   (a) Factor $f(x)$ into linear and irreducible quadratic factors with real coefficients.

   (b) Completely factor $f(x)$ into linear factors with complex coefficients.

8. Use $g(x) = 2x^5 - 3x^4 - 6x^3 + 14x^2 - 8x - 8$ to answer the questions below.

    (a) Factor $g(x)$ into linear and irreducible quadratic factors with real coefficients.

    (b) Completely factor $g(x)$ into linear factors with complex coefficients.

9. Refer to Example 10. Is $x = 3$ an upper bound for the zeros of $f(x) = 5x^3 - 15x^2 - 20x + 60$? Explain your answer.

For numbers 10 - 11, show all work and print the answer in the space provided.

10. According to the rational zeros theorem, which of the following is not a possible rational zero of $g(x) = 5x^4 - 6x^2 + 7x - 8$?

A.) 1   B.) $- 2$ C.) $- 2/5$

D.) $- 3/5$ E.) $1/5$

Answer:_____

11. According to the upper and lower bounds theorem, which of the following is not a lower bound of $f(x) = x^6 - 8x^5 + 17x^4 - 40x^3 + 64x^2 - 32x + 48$?

A.) -2   B.) -5   C.) -10   D.) 1

E.) According to the upper and lower bounds theorem, all are lower bounds.

Answer:_____

Chapter 20

Other types of Inequalities

20.1 Notes about Other types of Inequalities

In addition to linear inequalities, two other important types of inequalities are polynomial inequalities and rational inequalities. There are many similarities in solving these types of inequalities. The ideas listed below relate to solving these types of inequalities.

Idea 1 If a polynomial or rational expression is contained on one side of an inequality and 0 is on the other, then to solve the inequality it is sufficient to determine where the polynomial or rational expression is positive, negative and 0.

Idea 2 To make use of Idea 1, it is helpful to factor the polynomial or rational expression into linear terms.

Idea 3 Once the polynomial or rational expression has been factored into linear terms, determine where each of the terms is equal to 0. These points are called *critical numbers*. At each of the critical numbers, the expression is either 0 or undefined. At every other point, the expression is either positive or negative.

Idea 4 To determine the sign of the expression in the interval between each critical number, it is sufficient to find the value of the expression at one point in each interval.

Idea 5 If the inequality is not a strict inequality, then the critical numbers that make the original inequality true should be included. If the inequality is a strict inequality, no critical numbers should be included.

20.2 ❖ Errors to Avoid

Be careful not to treat inequalities as equations. Multiplying both sides of a
rational inequality by the same term can lead to errors. For example, the inequality
$\frac{x+1}{x-2} \geq 0$ has a solution of $(-\infty, -1] \cup (2, \infty)$, but $x + 1 \geq 0$ has a solution of
$[-1, \infty)$ (see Example 2.). Multiplying by $x - 2$ leads to an error.

20.3 Examples

EX 1. Find all real solutions of $x^2 + 3x < 4$.

> **Solution** ☞ Moving all terms to one side of the inequality and factoring
> the inequality into linear terms yields $x^2 + 3x < 4 \Rightarrow x^2 + 3x - 4 < 0 \Rightarrow$
> $(x + 4)(x - 1) < 0$. Determining the critical numbers gives $x + 4 = 0 \Rightarrow$
> $x = -4$ and $x - 1 = 0 \Rightarrow x = 1$. Thus, $x = -4$ and $x = 1$ are the critical
> numbers.
>
> Using the critical numbers to determine appropriate *check intervals* gives
> $(-\infty, -4)$, $(-4, 1)$, and $(1, \infty)$. Within each check interval, a *test point*
> will be chosen. Any point within the check intervals will suffice. For this
> example, we will use -6, 0 and 2 as test points.
>
> Substituting each test point into the non-zero side of factored inequality (in
> this case, the LHS) yields the following results:
>
> $((-6) + 4)((-6) - 1) = (-2)(-7) = 14 \Rightarrow 14 \not< 0 \Rightarrow -6$ does not make
> the inequality true and thus none of the points in $(-\infty, -4)$ is part of the
> solution.
>
> $((0) + 4)((0) - 1) = (-4)(1) = -4 \Rightarrow -4 < 0 \Rightarrow 0$ makes the inequality
> true and thus all of the points in the check interval that contained 0 are a
> solution. Moreover, $(-4, 1)$ must be part of the solution.
>
> $((2) + 4)((2) - 1) = (6)(1) = 6 \Rightarrow 6 \not< 0 \Rightarrow 2$ does not make the inequality
> true and thus none of the points in the check interval that contained 2 is a
> solution.
>
> Because $x^2 + 3x < 4$ contains a strict inequality, the endpoints of the check
> intervals (-4 and 1) cannot be solutions. Thus, the solution of the inequal-
> ity $x^2 + 3x < 4$ is $(-4, 1)$.

EX 2. Find all solutions of the inequality $\dfrac{x+1}{x-2} \geq 0$.

Setting each linear term equal to 0 yields $x + 1 = 0$ and $x - 2 = 0$, which have solutions $x = -1$ and $x = 2$, respectively. This gives check intervals of $(-\infty, -1), (-1, 2)$, and $(2, \infty)$. Choosing test points of -3, 0, and 5 gives that $\dfrac{x+1}{x-2}$ equals $2/5$, $-1/2$, and $3/2$, respectively. Because -1 makes the original inequality true and 2 makes the original inequality undefined, the solution of the inequality is $(-\infty, -1] \cup (2, \infty)$.

EX 3. Find all real solutions of $0 < (x - 2)(x^2 + 4x + 6)$.

Solution ☞ Using the quadratic formula to find the roots of $x^2 + 4x + 6 = 0$ shows that there are no real roots. Thus the only critical number is the solution of $x - 2 = 0$ or $x = 2$. Therefore the check intervals are $(-\infty, 2)$ and $(2, \infty)$. Choosing 0 and 3 as test points of $(x - 2)(x^2 + 4x + 6)$ gives values of -12 and 27, respectively. Because this is a strict inequality, the solution is $(2, \infty)$.

EX 4. ★ (S01#16) Solve the following inequality and express the solution in interval form.

$$\frac{3x+2}{x-6} \geq 4$$

Solution ☞ $\dfrac{3x+2}{x-6} \geq 4 \Rightarrow \dfrac{3x+2}{x-6} - 4 \geq 0 \Rightarrow \dfrac{3x+2-4(x-6)}{x-6} \geq 0 \Rightarrow$

$\Rightarrow \dfrac{3x+2-4x+24}{x-6} \geq 0 \Rightarrow \dfrac{-x+26}{x-6} \geq 0.$

The critical numbers of the inequality 6 and 26 divide the real line into three intervals $(-\infty, 6), (6, 26]$, and $[26, \infty)$. Note that we did not put a bracket around 6, since 6 makes the denominator zero. To determine where $\dfrac{-x+26}{x-6} \geq 0$, we pick a test point in each of the intervals.

Intervals	$(-\infty, 6)$	$(6, 26]$	$[26, \infty)$
Test points	$x = 0$	$x = 10$	$x = 27$
$\dfrac{-x+26}{x-6}$	$\dfrac{-0+26}{0-6} = -\dfrac{13}{3}$	$\dfrac{-10+26}{10-6} = 4$	$\dfrac{-27+26}{27-6} = -\dfrac{1}{21}$
Signs of $\dfrac{-x+26}{x-6}$	- negative	+ positive	- negative

Thus $\dfrac{-x+26}{x-6} \geq 0$, when x is in $(6, 26]$.

Therefore, the solution is $S = (6, 26]$.

EX 5.   ✭ (S03#12)Find all real solutions of $\dfrac{12}{x-5} \leq -2$.

Solution ☞ Adding 2 to both sides and factoring the inequality yields

$$\frac{12}{x-5} + 2 \leq 0 \Rightarrow \frac{12}{x-5} + \frac{2x-10}{x-5} \leq 0 \Rightarrow \frac{2x+2}{x-5} \leq 0.$$

Factoring each linear term to find the critical numbers gives $2x - 2 = 0 \Rightarrow$ $x = 1$ and $x - 5 = 0 \Rightarrow x = 5$. Thus the critical numbers are $x = 1, 5$ which means the check intervals are $(-\infty, -1) \cup (-1, 5) \cup (5, \infty)$. Picking -2, 0 and 10 as test points for $\dfrac{2x+2}{x-5}$ yields $4/7$, $-2/5$ and $22/5$, respectively. Because -1 makes the original inequality true and 5 makes the original inequality undefined, the solution of the inequality is $[-1, 5)$.

EX 6. Find all real solutions of $\dfrac{x^3 - 2x^2 + 2x + 4}{x+4} \geq 1$.

Solution ☞ Subtracting 1 from both sides of the inequality and factoring gives

$$\frac{x^3 - 2x^2 + 2x + 4}{x+4} \geq 1 \Rightarrow$$
$$\frac{x^3 - 2x^2 + 2x + 4}{x+4} - 1 \geq 0 \Rightarrow$$
$$\frac{x^3 - 2x^2 + 2x + 4}{x+4} - \frac{x+4}{x+4} \geq 0 \Rightarrow$$
$$\frac{x^3 - 2x^2 + x}{x+4} \geq 0 \Rightarrow$$
$$\frac{x(x^2 - 2x + 1)}{x+4} \geq 0 \Rightarrow \frac{x(x-1)^2}{x+4} \geq 0.$$

Finding the critical numbers by setting x, $(x-1)^2$, and $x+4$ equal to 0 yields 0, 1, and -4, respectively. This gives check intervals of $(-\infty, -4)$, $(-4, 0)$, $(0, 1)$, and $(1, \infty)$. Choosing -5, -1, $.5$, and 2 as test points gives the value of $\dfrac{x(x-1)^2}{x+4}$ as 180, $-4/3$, $\approx .0278$, and $1/3$, respectively. Because 0 and 1 make the original inequality true and -4 makes the original inequality undefined, the solution of the inequality is $(-\infty, -4) \cup [0, \infty)$.

Chapter 20 Exercises

For numbers 1 - 16, find all real solutions. Give your answer in interval form.

1. $(x-4)(x+3) \leq 0$

2. $0 > (x+7)(x+11)$

3. $(x+1)(x-2)(x-4) < 0$

4. $\dfrac{x+3}{x-4} \geq 0$

5. $0 \leq \dfrac{(x-5)(x+2)}{x+7}$

6. $\dfrac{(x-8)(x+6)}{(x+9)^2} > 0$

7. $0 \leq x^2 - 7x - 8$

8. $x^3 - 2x^2 + x > 0$

9. $\dfrac{x+1}{x^2-4} < 0$

10. $0 \leq \dfrac{x^2+3x+2}{x^2-4x-5}$

11. $\dfrac{x+7}{x-3} < 1$

12. $3 \geq \dfrac{x^2 + 4x}{x + 2}$

13. $\dfrac{x^2 - 4x - 5}{7 - x} \geq -2$

14. $\dfrac{-2x}{x + 1} + \dfrac{2x}{x^2 + 3x + 2} \leq \dfrac{2x^2}{x + 2}$

15. $\dfrac{x}{x-2} > \dfrac{12-3x}{x^2-4x+4}$

16. Two cyclists Yan and Clancy are in a race. Yan is better on shorter courses and Clancy is better on longer courses. Yan's average speed is given by $30 - x/10$ mph and Clancy's average speed is given by $28 - x/20$ mph, where x is the distance of the race in miles and $0 < x \le 200$. What distances could the race be for Clancy to win?

For numbers 17 - 19, show all work and print the answer in the space provided.

17. What are the critical numbers of the inequality $x^3 - 3x^2 + 2x \geq 0$.

   A.) $x = 0$ B.) $x = 0, 1, 2$ C.) $x = 0, -1, 2$

   D.) $x = -1, -2, 0$ E.) None of these.

   Answer:_____

18. What are the check intervals of $x^2 - 4x > 21$.

   A.) $(-\infty, 0), (0, 4),$ and $(4, \infty)$ B.) $(-\infty, -3), (-3, 7),$ and $(7, \infty)$

   C.) $(-\infty, 0), (0, 5.25),$ and $(5.25, \infty)$ D.) $(-\infty, -7), (-7, 3),$ and $(3, \infty)$

   E.) None of these.

   Answer:_____

19. Find all real solutions of the equation

$$\frac{1}{x+1} + \frac{2x}{x-2} < 2.$$

   A.) $(-\infty, -1) \cup (2, \infty)$ B.) $x = -2/5$ C.) $(-\infty, -1) \cup (-2/5, 2)$

   D.) $(-1, -2/5) \cup (2, \infty)$ E.) None of these.

   Answer:_____

Chapter 21

Variation

21.1   Notes about Variation

Direct Variation: If two quantities x and y are such that $y = kx$ (for some nonzero constant k), then x and y are said to vary directly or to be directly proportional.

Indirect Variation: If two quantities x and y are such that $y = \dfrac{k}{x}$ (for some nonzero constant k), then x and y are said to vary indirectly or to be indirectly proportional.

Joint Variation: If three quantities x, y and z are such that $z = kxy$ (for some nonzero constant k), then z is said to vary jointly with x and y or z is jointly proportional to x and y.

21.2   ❖ Errors to Avoid

Be aware that direct and indirect variation can be combined. See Example 2.

21.3 Examples

EX 1. Suppose that y varies directly as x and when $y = 50$, $x = 10$.

   (a) Find k.

   Solution ☞ Because y varies directly as x, we know that
   $y = kx \Rightarrow$
   $50 = k(10) \Rightarrow$
   $k = 5$.

   (b) If $y = 250$, what is x?

   Solution ☞ From the solution above, we know that $k = 5$. So,
   $250 = 5x \Rightarrow x = 50$.

   (c) If $x = 30$, what is y?

   Solution ☞ $y = kx \Rightarrow y = 5 \cdot 30 = 150$.

EX 2. ★ (S04#7) Which of the following represents a mathematical model of the statement F varies directly as g and inversely as r squared (here k represents a nonzero constant)?

A.) $F = kgr^2$ 　　　　 B.) $F = k\dfrac{g}{\sqrt{r}}$ 　　　　 C.) $F = kg\sqrt{r}$

D.) $F = k\dfrac{r^2}{g}$ 　　　　 E.) $F = k\dfrac{g}{r^2}$

Solution ☞ The problem describes $F = k\dfrac{g}{r^2}$.

The correct answer is **E**.

Name:_____

Section Number:_____

Date:_____

Chapter 21 Exercises

1. Suppose that y varies directly as x and when $y = 20$, $x = 30$.

   (a) Find k.

   (b) If $y = -10$, what is x?

   (c) If $x = 100$, what is y?

2. During a thunderstorm, you see lightning before you hear thunder. In particular, the distance between an observer and the lightning varies directly as the time difference between seeing and hearing the lightning and thunder, respectively.

   (a) Suppose an observer sees lightning hit a building she knows is 2,500 ft away. If it takes 4 seconds for her to hear the thunder, find the constant of proportionality.

   (b) Find the equation of variation. Include the value of k found in (a).

   (c) If you hear thunder 12 seconds after seeing lightning, how far away was the lightning strike?

   (d) If lightning strikes a building 6,000 ft away, how long will it take you to hear the thunder?

3. The number of CD's a particular band sells is inversely proportional to the price per CD.

(a) If the band sells 10,000 CD's when the price is \$8 per CD, find the constant of proportionality.

(b) Find the equation of variation. Include the value of k found in (a).

(c) If the band raises the price of the CD's to \$12.50 each, how many CD's will they sell?

(d) If the band sells 15,000 CD's, what price was charged per CD?

4. The monthly beer sales for a grocery store in a large city are directly proportional to the distance to the nearest liquor store and inversely proportional to the price charged for the beer (per case).

    (a) If the number of cases sold is 5,000 when the distance to the nearest liquor store is 1.2 miles and the price per case is \$15, find the constant of proportionality.

    (b) Find the equation of joint variation. Include the value of k found in (a).

    (c) If the store lowers its price to \$10 and the nearest liquor store is 1 mile away, how many cases will the store sell?

5. Which of the following represents a mathematical model of the statement M varies directly as l squared and inversely as the product of p and the square root of n (here k represents a nonzero constant)?

A.) $M = k\dfrac{l^2 p}{\sqrt{n}}$ B.) $M = k\dfrac{l^2 \sqrt{n}}{p}$ C.) $M = k\dfrac{l^2}{p\sqrt{n}}$

D.) $M = kl^2 p\sqrt{n}$ E.) $M = k\dfrac{p\sqrt{n}}{l^2}$

Answer:_____

Chapter 22

Exponential Functions

22.1 Notes about Exponential Functions

Definition: An exponential function is a function of the form $f(x) = a^x$, where x is any real number, a is called the base, $a > 0$ and $a \neq 1$.

The graph of an exponential function $f(x) = a^x$, where $a > 1$ has no x-intercept, a y-intercept at $(0, 1)$, has a horizontal asymptote at $y = 0$ and the graph is shown in Figure 22.1.

The graph of an exponential function $f(x) = a^x$, where $0 < a < 1$ has no x-intercept, a y-intercept at $(0, 1)$, has a horizontal asymptote at $y = 0$ and the graph is shown in Figure 22.2.

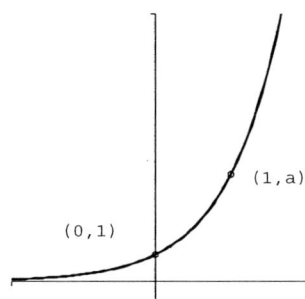

Figure 22.1: Graph of $f(x) = a^x$, $a > 1$.

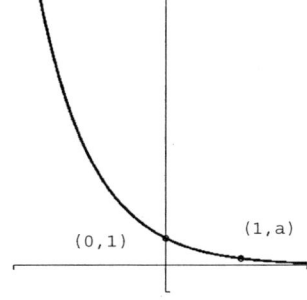

Figure 22.2: Graph of $f(x) = a^x$, $0 < a < 1$.

22.2 ❖ Errors to Avoid

While all exponential functions of the form $f(x) = a^x$ with $a > 1$ have the same basic shape, as a grows larger, the graph will grow more steeply as the x values increase. Note the difference in the graphs of $f(x) = 2^x$ (solid) and $g(x) = 5^x$ (dashed) below.

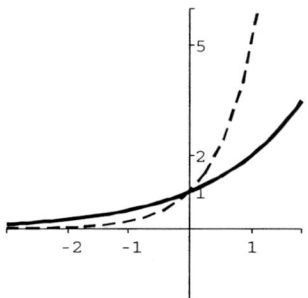

Figure 22.3: Graph of $f(x) = 2^x$ (solid), $g(x) = 5^x$ (dashed).

22.3 Examples

EX 1. Sketch the graphs of $f(x) = 3^x$ and $g(x) = \left(\dfrac{2}{5}\right)^x$.

Solution ☞ To sketch the graph of $f(x) = 3^x$, we note that this is an exponential function with $a > 1$. The graph should include the points $(0, 1)$ and $(1, 3)$. The graph is sketched in Figure 22.4.

To sketch the graph of $g(x) = \left(\dfrac{2}{5}\right)^x$, we note that this is an exponential function with $0 < a < 1$. The graph should include the points $(0, 1)$ and $(1, 2/5)$. The graph is sketched in Figure 22.5.

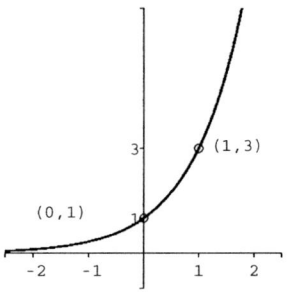

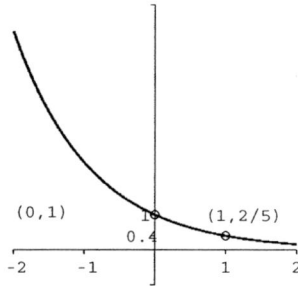

Figure 22.4: Graph of $f(x) = 3^x$.

Figure 22.5: Graph of $g(x) = \left(\dfrac{2}{5}\right)^x$.

EX 2. Sketch the graph of $h(x) = 2^{x+4} - 1$.

Solution ☞ Because we know what the graph of $y = 2^x$ looks like, we can sketch the graph of $h(x)$ by moving the graph left 4 units and down 1 unit. To make this easier, we will first move individual points from the graph of $y = 2^x$. Since $(0, 1)$, $(1, 2)$, and $(2, 4)$ are on the graph of $y = 2^x$, then $(-4, 0)$, $(-3, 1)$, and $(-2, 3)$ should be on the graph of $h(x) = 2^{x+4} - 1$. The horizontal asymptote should also be moved down 1 unit, giving the graph in Figure 22.7.

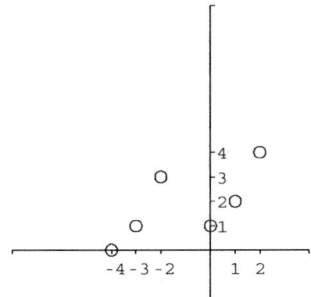

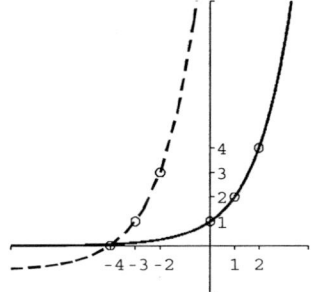

Figure 22.6: Graph of points plotted.

Figure 22.7: Graph of $h(x) = 2^{x+4} - 1$ (dashed) and $y = 2^x$ (solid).

EX 3. The graph below is of the form $f(x) = C \cdot a^x$. Find C and a.

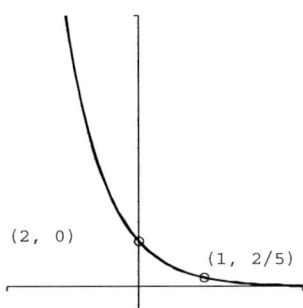

Figure 22.8: $y = C \cdot a^x$.

Solution ☞ To find C, we notice that $f(0) = C \cdot a^0 = C \cdot 1 = C$. From the graph, we see that when $x = 0$, $y = 2$. Thus, $C = 2$.
Now using the fact that $f(1) = 2/5 = 2 \cdot (a^1)$. Dividing both sides by 2 gives $a = 1/5$.

EX 4. If \$1000 is invested at 6% compounded monthly, then the amount A , in dollars, in the bank after t months is given by $A(t) = 1000 \cdot (1.005)^t$.

(a) How much money will be in the bank after 3 years?

Solution ☞ Since $A(t)$ represents the amount of money, we must find $A(36)$ (because t is measured in months). We find that $A(36) = 1000 \cdot (1.005)^{36} \approx \$1196.68$.

(b) Sketch the graph of $y = A(t)$ over 25 years.

Solution ☞ The graph is below. The horizontal axis goes to 300 because 25 years is equivalent to 300 months. Note that because a is close to 1, the graph is not very steep.

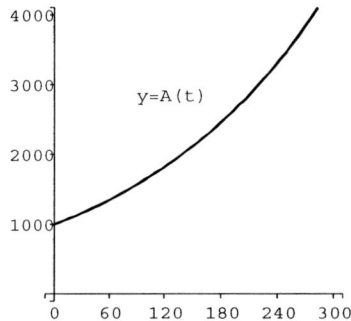

(c) Estimate when $A(t)$ will equal \$3000 to the nearest year.

Solution ☞ To estimate this, we will graph $y = 3000$ along with the graph of $A(t) = 1000 \cdot (1.005)^t$. Doing this, its seems that the value for t should be between $t = 200$ and $t = 240$. Graphing over a smaller interval yields Figure 22.10. We see that \$3000 is reached between the 216th and 228th month or between the 18th and 19th years, thus the answer is 19.

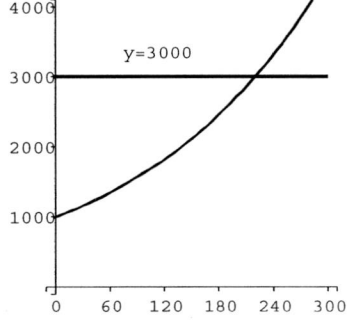

Figure 22.9: Graphs of $y = A(t)$ and $y = 3000$.

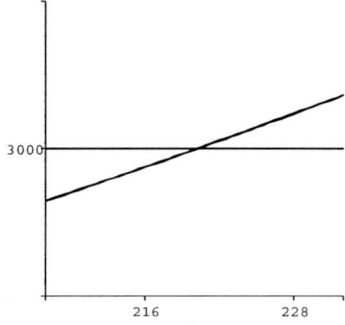

Figure 22.10: Graphs of $y = A(t)$ and $y = 3000$ zoomed in.

Using methods developed in Chapter 22, you will be able to solve this problem exactly.

Name:_____

Section Number:_____

Date:_____

Chapter 22 Exercises

Evaluate the given exponential functions at the given value for problems 1 - 4. Round the answers to three decimal places.

1. $f(x) = 4^x$ at $x = 2.5$

2. $f(z) = 3^z$ at $z = 7$

3. $g(s) = 0.12^s$ at $s = 4$

4. $h(t) = 5.7^t$ at $t = \sqrt{3}$

5. Explain how you could find the answer to 1 without using a calculator.

6. $\sqrt{3} \approx 1.73205080757$, but because $\sqrt{3}$ is an irrational number the decimal approximation never ends. Explain in your own words why this means that $2^{\sqrt{3}}$ is between 2^1 and 2^2. Further, explain how you could find an approximation of $2^{\sqrt{3}}$.

For problems 7 - 9, make a table of values for each function, plot the points on the graph and sketch the graph of the function.

7. $f(x) = \left(\dfrac{1}{3}\right)^x$.

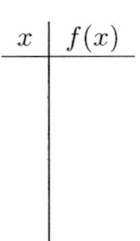

x	$f(x)$

8. $g(x) = 2^x$.

x	$g(x)$

9. $h(x) = \left(\dfrac{1}{2}\right)^{x+2} - 3.$

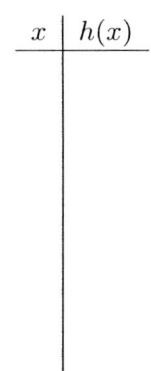

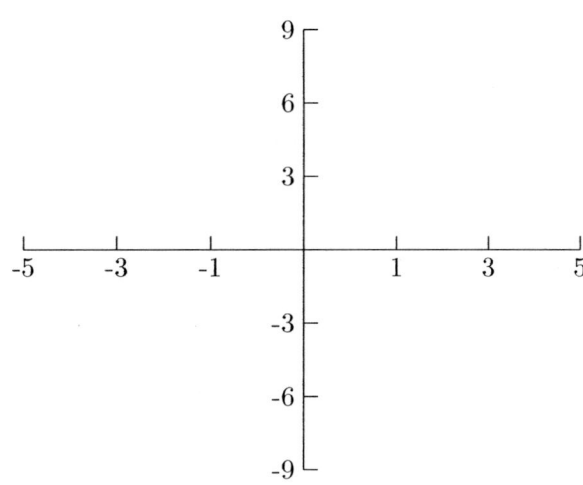

10. Anestra invests $5000 into a bank account that pays 4% interest compounded monthly. If she leaves the money in the bank for 7 years, how much money will be in the bank?

11. In number 10, how much interest did Anestra earn?

12. Salli and Emeriarta are saving to buy a condiminium. The down payment for the condo is $25000. How much money should the couple invest in the bank now to be able to make the down payment in 7 years if the interest rate at the bank is 8% compounded quarterly?

13. The population of College Township is given by the function $P(t) = 19960e^{.035t}$, where P represents the number of people and t represents time (in years) since 2001.

    a) How many people lived in College Township in 2001?

    b) How many people will live in College Township in 2007 (round your answer down to the nearest person)?

14. A certain type of radioactive material is decaying. The amount of the material remaining A (measured in μg) is determined by the function $A(t) = 125e^{-.12t}$, where t represents the time in hours since 2pm today. Determine how much of the material remains at 6:30pm today.

For numbers 15 - 18, show all work and print the answer in the space provided.

15. How is the graph of $g(x) = 4^{x-3} + 2$ shifted from the graph of $f(x) = 4^x$?

    A.) Left 2, up 3 units.   B.) Right 2, down 3 units.

    C.) Right 3, up 2 units.   D.) Left 3, up 2 units.

    E.) None of these.

Answer:_____

16. Which of the following would represent the graph of $y = 2^x$ shifted down 7 units and right 5 units?

    A.) $f(x) = 2^{(x-7)} + 5$ B.) $f(x) = 2^{(x-5)} - 7$ C.) $f(x) = 2^{(x+5)} - 7$

    D.) $f(x) = 2^{(x+7)} - 5$ E.) None of these.

Answer:_____

223

17. What is the domain and range of $g(x) = 5^{x-4} + 2$?

    A.) Domain: $x \geq 4$, Range: $y > -2$ B.) Domain: $x > 2$, Range: $y \geq 4$

    C.) Domain: All reals, Range: $y > 2$ D.) Domain: All reals, Range: $y > -2$

    E.) None of these.

                                              Answer:_____

18. The graph below is of the form $f(x) = 5^{x-a} + b$. Find a and b.

    A.) $a = 3$ and $b = 4$ B.) $a = -3$ and $b = 4$ C.) $a = 4$ and $b = -3$

    D.) $a = -3$ and $b = -4$ E.) None of these.

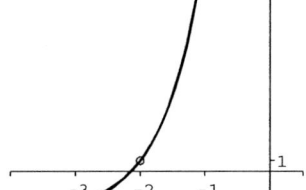

                                              Answer:_____

Chapter 23

Logarithmic functions

23.1 Notes about Logarithmic functions

For any real number $a > 0$, $a \neq 1$, the exponential function $f(x) = a^x$ is one-to-one and therefore has an inverse. This inverse function is called the *logarithmic function* with base a. It is denoted by $\log_a(x)$ and read "log base a of x".

Definition: If $a > 0$ and $a \neq 1$, then the logarithmic function with base a is the function $f(x) = \log_a(x)$, where $x > 0$ and $y = \log_a(x) \iff a^y = x$.

The domain of the logarithmic function is the range of the exponential function $(0, \infty)$. The range of the logarithmic function is the domain of the exponential function $(-\infty, \infty)$.

Note: $\log_a(x)$ represents the *power that a must be raised to to get x*.

Definition: The logarithmic function with base 10 is called the *common logarithm* and is denoted $f(x) = \log(x)$.

Definition: The logarithmic function with base e is called the *natural logarithm* and is denoted $f(x) = \ln(x)$.

The graph of a logarithmic function $f(x) = \log_a(x)$, where $a > 1$ has an x-intercept at $(1, 0)$, no y-intercept, has a vertical asymptote at $x = 0$ and the graph is shown in Figure 23.1.

The graph of an logarithmic function $f(x) = \log_a(x)$, where $0 < a < 1$ has an x-intercept at $(1, 0)$, no y-intercept, has a vertical asymptote at $x = 0$ and the graph is shown in Figure 23.2.

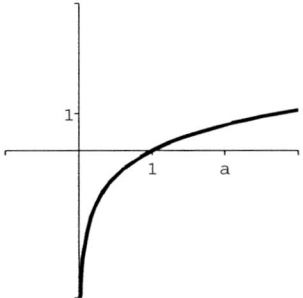   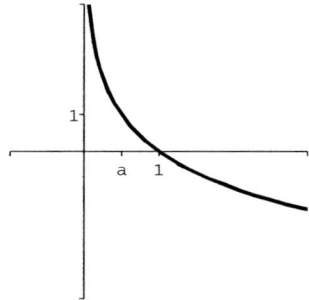

Figure 23.1: Graph of $f(x) =$ Figure 23.2: Graph of $f(x) =$
$\log_a(x)$, $a > 1$ $\log_a(x)$, $0 < a < 1$

23.2 Examples

EX 1. Sketch the graph of $y = \log_2(x)$.

Solution ☞ First, we will determine some points on the graph by using the idea that $\log_2(x)$ represents the power that x must be raised to to give 2. So, $x = 2 \Rightarrow \log_2(x) = 1$, $x = 4 \Rightarrow \log_2(x) = 2$, etc. Several other values are given in the table below.

x	y
1/4	-2
1/2	-1
1	0
2	1
4	2
8	3

$\Rightarrow$

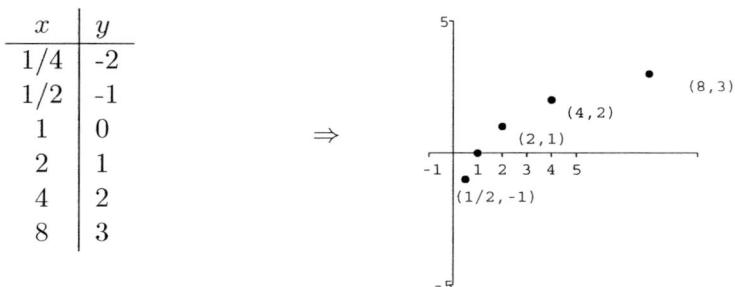

Connecting the points and using our knowledge of the shape of a logarithmic graph with $a > 1$ gives

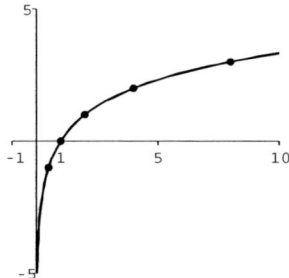

Note that we could have used the fact that $y = \log_2(x)$ is the inverse of the exponential function $y = 2^x$ and obtain the graph of $y = \log_2(x)$ by reflecting the graph of $y = 2^x$ across the line $y = x$.

EX 2. Find the domain of the function $f(x) = \log_3(x + 5)$.

Solution ☞ The domain of f is the set of all real numbers x for which $x + 5 > 0 \Rightarrow x > -5$. Therefore the domain is $(-5, \infty)$.

EX 3. Find the domain of the function $g(x) = \dfrac{\ln(x+7)}{x-2}$.

Solution ☞ The domain of g is all real numbers x for which $x + 7 > 0$ and $x - 2 \neq 0$. That is, all real numbers $x > -7$ and $x \neq 2$, which graphed below.

Thus the domain is $(-7, 2) \cup (2, \infty)$.

EX 4. Sketch the graph of $y = \log_2(x+1) + 3$.

Solution ☞ The graph of $y = \log_2(x+1) + 3$ is the graph of $y = \log_2(x)$ shifted 1 unit to the left and up 3 units up.

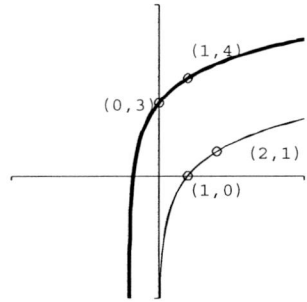

Figure 23.3: Graph of $y = \log_2(x+1) + 3$ (thick) and $y = \log_2(x)$ (thin).

EX 5. Find the exact value of $\log_2(1/8)$.

Solution ☞ To find this, recall that $\log_a(x)$ represents the exponent to which a must be raised to equal x. So we want to know what power 2 must be raised to to yield $1/8$. Because $1/8 = 8^{-1} = (2^3)^{-1} = 2^{-3}$, $\log_2(1/8) = -3$.

EX 6. Write the logarithmic equation $\log_k(m-1) = b$ in exponential form.

Solution ☞ Using the fact that $y = \log_a(x) \iff x = a^y$ with $y = b$, $a = k$ and $x = m - 1$ to get $m - 1 = k^b$.

EX 7. Write the exponential equation $e^{kt} = m$ in logarithmic form.

Solution ☞ Again, we use $y = \log_a(x) \iff x = a^y$ with $y = kt$, $a = e$ and $x = m \Rightarrow kt = \log_e m$ or $kt = \ln m$.

EX 8. Find the exact value of $\ln\left(\dfrac{1}{\sqrt{e}}\right)$.

Solution ☞ $\dfrac{1}{\sqrt{e}} = \dfrac{1}{e^{1/2}} = e^{-1/2}$, so $\ln\left(\dfrac{1}{\sqrt{e}}\right) = \ln e^{-1/2} = -1/2$ (since $-1/2$ is the power we should raise e to to give $e^{-1/2}$).

EX 9. For what base a does $\log_a\left(\dfrac{1}{81}\right) = 4$.

Solution ☞ If we write $\log_a\left(\dfrac{1}{81}\right) = 4$ in exponential form, we get $a^4 = \dfrac{1}{81}$. Note that $\dfrac{1}{81} = \dfrac{1}{3^4} = \left(\dfrac{1}{3}\right)^4 \therefore a^4 = \left(\dfrac{1}{3}\right)^4$ and $a = \dfrac{1}{3}$. Note that $a \neq -\dfrac{1}{3}$ because by definition $a > 0$.

Chapter 23 Exercises

For problems 1 - 6, find the exact value of the given logarithm.

1. $\log_2 16$

2. $\ln \sqrt{e}$

3. $\log 1000$

4. $\log_3 \dfrac{1}{9}$

5. $\log_a \dfrac{1}{a^5}$

6. $\ln e$

7. Find the domain of the function $f(x) = \ln(2x - 3)$.

8. Find the domain of the function $g(x) = \dfrac{\log(2x + 7)}{x - 3}$.

9. Sketch the graph of the function $y = \log_2(x)$. Make sure to include the x- and y-intercepts, as well as at least two other points.

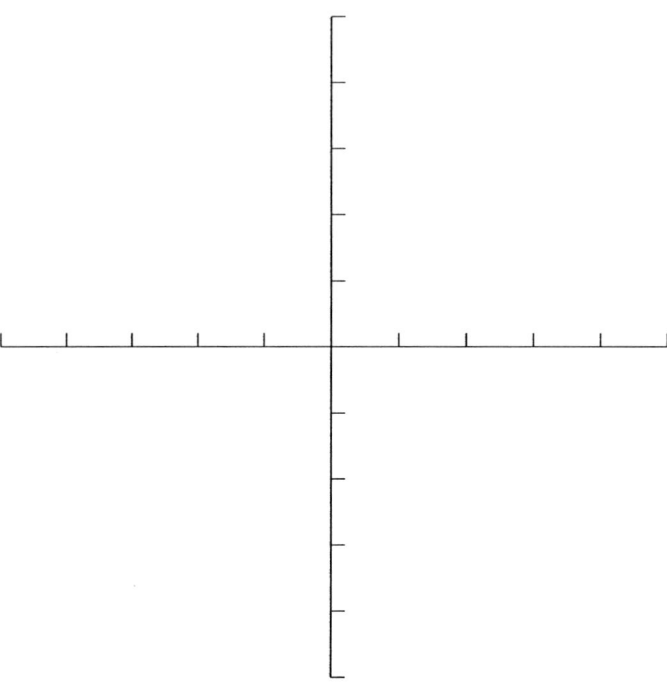

10. Sketch the graph of the function $y = \log_{1/3}(x)$. Make sure to include the x- and y-intercepts, as well as at least two other points.

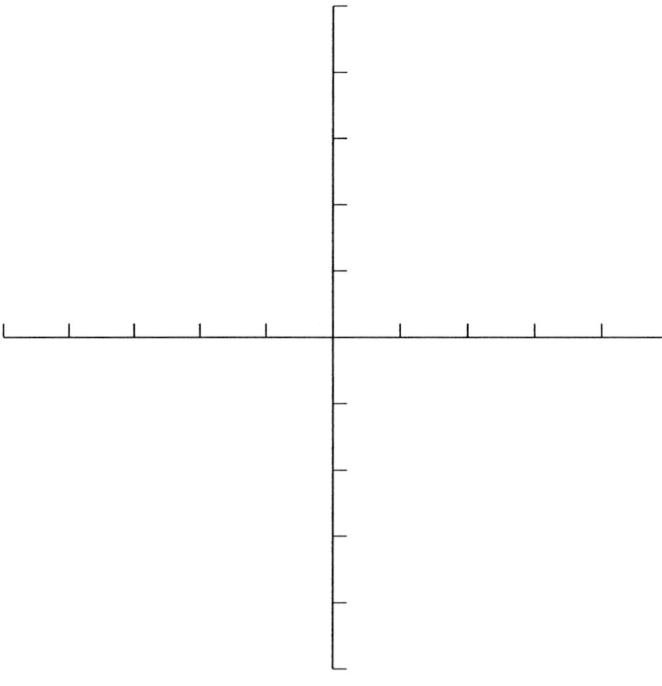

11. Sketch the graph of the function $y = \ln(x - 1)$. Make sure to include the x- and y-intercepts, as well as at least two other points.

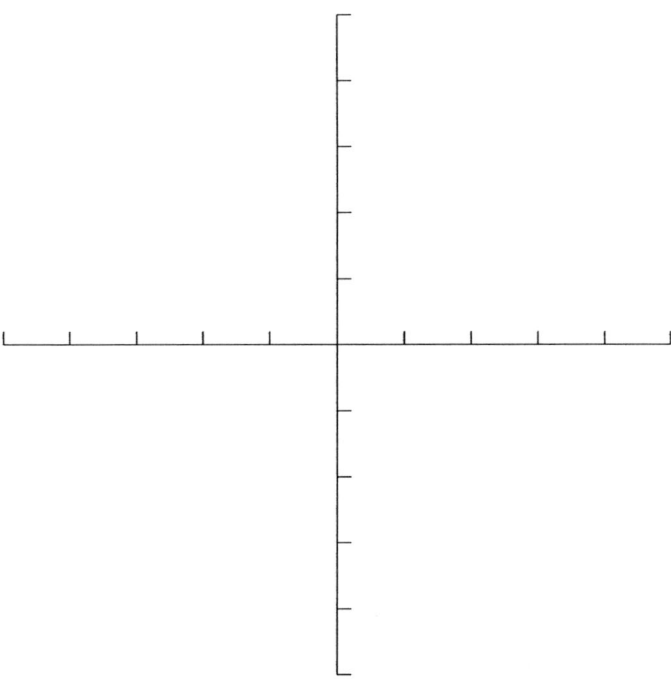

12. Sketch the graph of the function $y = -\log(x + 2)$. Make sure to include the x- and y-intercepts, as well as at least two other points.

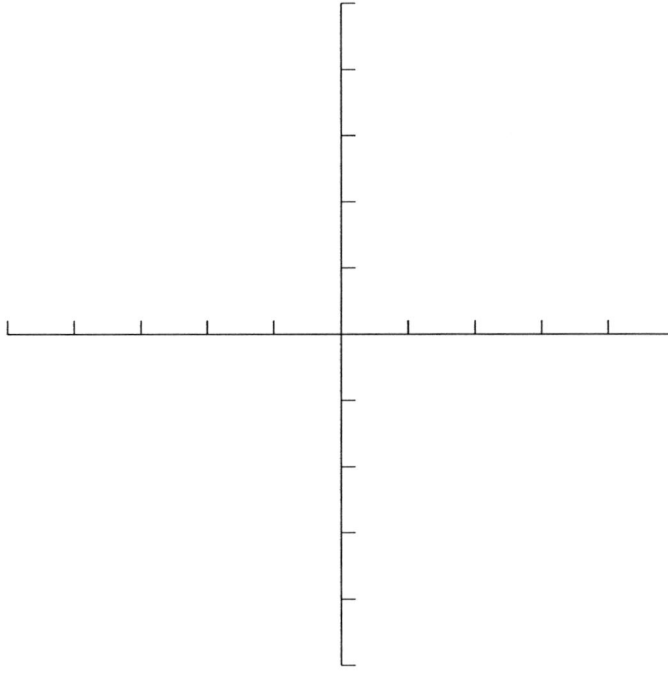

13. Rewrite the exponential expression $b^n = k$ in logarithmic form.

14. Rewrite the logarithmic expression $\log_k(x + 5) = 3$ in exponential form.

15. For what value of x is $\log_2(x) = 3$?

16. For what value of x is $\log(x) = -1$?

17. For what value of x is $\ln(x) = \frac{1}{4}$?

For numbers 18 - 22, show all work and print the answer in the space provided.

18. $\log_a a^5 =$

A.) 1   B.) a C.) a^5 D.) 5   E.) $\log_a 5$

Answer:_____

19. $\log_2 32 =$

A.) 16   B.) 5   C.) 4   D.) 2   E.) 1

Answer:_____

20. $\log_{1/2} 8 =$

A.) -3 B.) 3   C.) $2\sqrt{2}$ D.) 4   E.) -4

Answer:_____

21. Write the logarithmic equation $\log_k(x-1) = b$ in exponential form.

A.) $x - 1 = b^k$ B.) $(x-1)^k = b$ C.) $k = (x-1)^b$
D.) $k(x-1) = b$ E.) $x - 1 = k^b$

Answer:_____

22. What is the value of the base a for which $\log_a 64 = 3$.

A.) 10   B.) 8   C.) 4   D.) e E.) 2

Answer:_____

Chapter 24

Properties and Laws of Logarithmic Functions

24.1   Notes about Logarithmic Functions

Definition: The *logarithmic function* with base $a, (a > 0, a \neq 1)$ is

$$f(x) = \log_a x$$

where $\log_a x$ is defined by $\log_a x = y \Leftrightarrow x = a^y$.

In other words, $\log_a x$ is the power to which the base a must be raised to get x.

Important: It is necessary to know how to use the property
$\log_a x = y \Leftrightarrow x = a^y$ to rewrite a logarithmic expression or equation in exponential form and vis versa.

Some Basic Properties of Logarithmic Functions:

i. $\log_a 1 = 0$, since $a^0 = 1$.

ii. $\log_a a = 1$, since $a^1 = a$.

iii. $\log_a a^x = x$, since $a^x = a^x$.

iv. $a^{\log_a x} = x$, by definition.

Laws of Logarithms:

Let $a > 0$, $a \neq 1$, $M > 0$, $N > 0$, and r be any real number, then

1. $\log_a(M \cdot N) = \log_a(M) + \log_a(N)$

2. $\log_a\left(\dfrac{M}{N}\right) = \log_a(M) - \log_a(N)$

3. $\log_a M^r = r \cdot \log_a M$

The above properties and laws are true for all bases. In particular for base 10 and base e. Recall that a logarithm with base 10 is called *common logarithm*, it is denoted by log and a logarithm with base e is called *natural logarithm* it is denoted by ln.

Change of Base Formula: If x, a and b are positive numbers, with $a \neq 1$ and $b \neq 1$, then

$$\log_a x = \frac{\log_b x}{\log_b a}.$$

Many calculators will evaluate common logarithm and natural logarithms. The change of base formula allows you to use a calculator to find the value of a logarithm of base other than 10 or e (see Example 7).

24.2   ❖ Errors to Avoid

Below are some common mistakes. Be careful to avoid them.

$$\log_a(M + N) \neq \log_a(M) + \log_a(N)$$

$$\log_a\left(\frac{M}{N}\right) \neq \frac{\log_a(M)}{\log_a(N)}$$

$$\left(\log_a M\right)^r \neq r \log_a M$$

24.3   Examples

EX 1. Express the logarithmic equation $\log_4(3x - 1) = 2$ in exponential form

   Solution ☞ $3x - 1 = 4^2$.

EX 2. Write the logarithmic equation $\log_b(t^2 + 5) = s - 7$ in exponential form

   Solution ☞ $t^2 + 5 = b^{s-7}$.

EX 3. Express the exponential equation $e^{x-10} = 9$ in logarithmic form

   Solution ☞ $\log_e 9 = x - 10$ or $\ln 9 = x - 10$.

EX 4. Use the Laws of logarithms to write the expression $\log_2 \dfrac{\sqrt{x}}{y^3 z}$ in a form with no logarithm of product, quotient, or power.

   Solution ☞ Again using the second law of logarithms, we see that

$$\log_2 \frac{\sqrt{x}}{y^3 z} = \log_2 \sqrt{x} - \log_2 y^3 z$$
$$= \log_2 x^{\frac{1}{2}} - (\log_2 y^3 + \log_2 z)$$
$$= \frac{1}{2} \log_2 x - 3 \log_2 y - \log_2 z.$$

EX 5. Use the Laws of logarithms to write the expression

$\dfrac{1}{3} \log_7(x+2) - 2 \log_7(y-1) - \log_7(z+4)$ as a single logarithm.

   Solution ☞ Now we use laws of logarithms in the other direction.

$$\frac{1}{3} \log_7(x+2) - 2\log_7(y-1) - \log_7(z+4) \qquad =$$
$$\log_7(x+2)^{\frac{1}{3}} - \left(\log_7(y-1)^2 + \log_7(z+4) \right) \qquad =$$
$$\log_7 \sqrt[3]{x+2} - \log_7 (y-1)^2(z+4) \qquad = \log_7 \frac{\sqrt[3]{x+2}}{(y-1)^2(z+4)}.$$

EX 6. Use the Laws of logarithms to write the expression

$\ln 9 + 2 \ln \dfrac{x-1}{3} - 4 \ln \sqrt{2x+5}$ as a single logarithm.

   Solution ☞ Again we use laws of logarithms in the other direction.

$$\ln 9 + 2 \ln \frac{x-1}{3} - 4 \ln \sqrt{2x+5} = \ln 9 + \ln \left(\frac{x-1}{3} \right)^2 - \ln \left(\sqrt{2x+5} \right)^4$$
$$= \ln \left(9 \cdot \frac{(x-1)^2}{9} \right) - \ln(2x+5)^2$$
$$= \ln(x-1)^2 - \ln(2x+5)^2$$
$$= \ln \frac{(x-1)^2}{(2x+5)^2}.$$

EX 7. Use the Change of Base Formula and a calculator to evaluate $\log_5 1521$. Round your answer to four decimal places.

Solution ☞ Using the Change of Base Formula we can say that

$$\log_5 1521 = \frac{\ln 1521}{\ln 5} \approx 4.5526.$$

EX 8. Show that $\log(10^x + 1) = x + \log(10^{-x} + 1)$.

Solution ☞ To show that $\log(10^x + 1) = x + \log(10^{-x} + 1)$, we prove $\log(10^x + 1) - \log(10^{-x} + 1) = x$.

$$\log(10^x + 1) - \log(10^{-x} + 1) = \log(10^x + 1) - \log\left(\frac{1}{10^x} + 1\right)$$

$$= \log(10^x + 1) - \log\left(\frac{1 + 10^x}{10^x}\right)$$

$$= \log\left(\frac{10^x + 1}{\frac{1+10^x}{10^x}}\right)$$

$$= \log 10^x = x.$$

EX 9. ☆ (S02#14) Which of the following is $3\ln(2x+1)+1/2[\ln(x-4)-\ln(x^4+1)]$ correctly written as a single logarithm?

A.) $\ln \dfrac{3(2x+1)(x-4)}{2(x^4+1)}$

B.) $\ln \dfrac{(2x+1)^3 \sqrt{x-4}}{\sqrt{x^4+1}}$

C.) $\dfrac{\ln(2x+1)^3 \sqrt{x-4}}{\ln(\sqrt{x^4+1})}$

D.) $\log \dfrac{(2x+1)^3 \sqrt{x-4}}{\sqrt{x^4+1}}$

E.) $\dfrac{3}{2}\ln \dfrac{(2x+1)(x-4)}{(x^4+1)}$

Solution ☞

$$3\ln(2x+1)+1/2[\ln(x-4)-\ln(x^4+1)]=$$

$$\ln(2x+1)^3 + 1/2[\ln \dfrac{x-4}{x^4+1}] \qquad =$$

$$\ln(2x+1)^3 + \ln\left(\dfrac{x-4}{x^4+1}\right)^{1/2} \qquad =$$

$$\ln\left((2x+1)^3 \cdot \left(\dfrac{x-4}{x^4+1}\right)^{1/2}\right) \qquad = \ln \dfrac{(2x+1)^3 \sqrt{x-4}}{\sqrt{x^4+1}}.$$

EX 10. Write y as a function of x if $\ln y = 3\ln(x-1) + 2\ln x + \ln C$, where C is a constant.

Solution ☞ First we use laws of logarithms to write the right hand side of the expression as a single logarithm. Then we use the property that says if $\ln M = \ln N$, then $M = N$.

$$\ln y = 3\ln(x-1) + 2\ln x + \ln C = \ln(x-1)^3 + \ln x^2 + \ln C$$
$$= \ln(x-1)^3 x^2 + \ln C$$
$$= \ln C(x-1)^3 x^2$$

Hence, $y = C(x-1)^3 x^2$.

EX 11. Write y as a function of x if $\ln y = 5x + \ln C$, where C is a constant.

Solution ☞ If we write $\ln y = 5x + \ln C$ in exponential form we get
$y = e^{5x + \ln C} = e^{5x} e^{\ln C} = Ce^{5x}$.

Therefore, $y = Ce^{5x}$.

Name:_____

Section Number:_____

Date:_____

Chapter 24 Exercises

Use the Laws of logarithms for problems 1 - 4 to write the expression in a form with no logarithm of product, quotient, or power.

1. $\log_5 \dfrac{x^3 \sqrt{y}}{z^5}$

2. $\log \dfrac{x(x^3+1)^3}{\sqrt{x^2+3}}$

3. $\ln \sqrt{\dfrac{3e^{x+1}}{(x^2+5)(x-6)^3}}$

4. $\log_3 \dfrac{x^2 \sqrt{x^2+1}}{\sqrt[3]{x-6}}$

Use the Laws of logarithms to write each of numbers 5 - 8 as a single logarithm.

5. $3\log(x-2) + \log(y-1) - \dfrac{1}{3}\log(z+1)$

6. $\dfrac{2}{3}\ln(2x+1) - \ln(y-1) + 3\ln(z+4)$

7. $2\left[\log_3 x + \dfrac{1}{2}\log_3(x+2) - \log_3(x-5)\right]$

8. $\log_8(x^2-9) + 2\log_8 x - \log_8(x+3)$

For numbers 9 - 15, show all work and print the answer in the space provided.

9. Write $\log_3(x - 1) = 4$ in exponential form.

   A.) $(x - 1)^3 = 4$ B.) $3^{x-1} = 4$ C.) $(x - 1) = 3^4$

   D.) $(x - 1) = 4^3$ E.) None of these.

Answer:_____

10. Write $\ln(m + 1) = n$ in exponential form.

   A.) $m + 1 = e^n$ B.) $n = e^{m+1}$ C.) $m + 1 = n^e$

   D.) $n + 1 = e^m$ E.) None of these.

Answer:_____

11. Write $k^{x-1} = y$ in logarithmic form.

   A.) $\log_k (x - 1) = y$ B.) $\log_k y = x - 1$ C.) $\log_y k = x - 1$

   D.) $\log_{x-1} k = y$ E.) None of these.

Answer:_____

12. If $\log_a M = 4$ and $\log_a N = 3$ then $\log_a \left(\dfrac{M}{N} \right) =$

   A.) $\dfrac{4}{3}$ B.) 1   C.) 7

   D.) 12   E.) None of these.

Answer:_____

13. Use the Change of Base Formula and a calculator to evaluate $\log_7 2141$. Round your answer to four decimal places.

A.) 1.0956   B.) 3.9411   C.) 5.7231

D.) 7.6658   E.) None of these.

Answer:_____

14. Rewrite the expression $5 \log x - \log y - \log(z+1)$ as the logarithm of a single quantity.

A.) $\log \dfrac{x^5}{y(z+1)}$ B.) $\log(x^5 y - z + 1)$ C.) $\log \dfrac{x^5 y}{z+1}$

D.) $\log(x^5 + y - z - 1)$ E.) None of these.

Answer:_____

15. $\log_a a^4 = ?$

A.) a B.) a^4 C.) 1   D.) 4   E.) None of these.

Answer:_____

Chapter 25

Exponential and Logarithmic Equations

25.1 Exponential and Logarithmic Equations

Definition: An *exponential equation* is an equation in which the variable occurs in the exponent.

Definition: A *logarithmic equation* is an equation in which the variable occurs in the logarithm.

Important: To solve exponential and logarithmic equations, we will use the following properties $(a > 0)$:

i. $\log_a x = y \Leftrightarrow x = a^y$

ii. $\log_a M = \log_a N \Rightarrow M = N$

iii. $a^M = a^N \Rightarrow M = N$

iv. If $M > 0, N > 0$ and $M = N$, then $\log_a M = \log_a N$

We solve logarithmic equations by using the following steps:

1. Take all the logarithmic terms to the left hand side of the equation.

2. Rewrite the left hand side of the equation as a single logarithm, using the Laws of logarithms as necessary.

3. Write the equation in its exponential form and solve for the variable.

4. Check your solutions.

To solve exponential equations we use the following steps:

1. Isolate the exponential term to one side of the equation.

2. Take the logarithm of both sides of the equation, then use the third Law of Logarithms (see the previous chapter) to bring down the exponent.

3. Solve the resulting equation for the variable.

25.2 ❖ Errors to Avoid

In order to solve some exponential equations like $9^x - 3^{x+1} = 28$, the above procedure is not sufficient. Some manipulation of the equation is required before we can apply the above steps. See Example 9 for the solution.

To avoid incorrectly yielding solutions to logarithmic equations that are not in the domain of the original function, all potential solutions should be checked.

25.3 Examples

EX 1. Solve the logarithmic equation $\log_4(3x - 5) = 2$.

Solution ☞ The first two steps of solving logarithmic equations are already done for this example. So all that we need to do is to write the equation in its exponential form, solve for the variable, and then check our solutions. $\log_4(3x - 5) = 2 \Rightarrow 3x - 5 = 4^2 \Rightarrow 3x = 16 + 5 \Rightarrow x = 7$.

✎ To check our answer, we replace x with 7 in the original equation $\log_4(3x - 5) = 2$. Obtaining $\log_4(3 \cdot 7 - 5) = \log_4 16 = 2$ and our solution is correct.

EX 2. Solve the logarithmic equation $\log_2 2x + \log_2(x - 3) = 3$.

Solution ☞ Using the first law of logarithm, we can say $\log_2 2x + \log_2(x - 3) = 3 \Rightarrow \log_2 2x(x - 3) = 3$.

Now we can write the equation in its exponential form $2x(x - 3) = 2^3 \Rightarrow 2x^2 - 6x = 8 \Rightarrow x^2 - 3x - 4 = 0$.

This is a quadratic equation in x. We can solve this equation by factoring. $x^2 - 3x - 4 = 0 \Rightarrow (x + 1)(x - 4) = 0 \Rightarrow x = -1$ or $x = 4$.

✎ To check our solutions, we first replace x with -1 in the original equation $\log_2 2x + \log_2(x - 3) = 3$. However, -1 makes the left hand side of the equation undefined. Therefore, -1 is not a solution.

Next replace x with 4 in the original equation, obtaining $\log_2 8 + \log_2 1 = 3 - 0 = 3$ and the solution is $x = 4$.

EX 3. Solve the logarithmic equation $\ln(2x+1) - \ln(x-2) = 1$.

Solution ☞ Using the second law of logarithms, we can say
$$\ln(2x+1) - \ln(x-2) = 1 \Rightarrow \ln\frac{2x+1}{x-2} = 1.$$

Changing the equation to its exponential form gives
$$\frac{2x+1}{x-2} = e^1 \Rightarrow 2x+1 = e(x-2).$$

This is a linear equation in x. We can solve it by taking all the terms that involve x to one side and the constant to the other side.

$$2x+1 = ex - 2e \Rightarrow 1 + 2e = ex - 2x \Rightarrow 1 + 2e = (e-2)x \Rightarrow x = \frac{1+2e}{e-2}.$$

To check our solution, replace x with $\frac{1+2e}{e-2}$ in the original equation. We have

$$\ln(2\frac{1+2e}{e-2} + 1) - \ln(\frac{1+2e}{e-2} - 2) =$$
$$\ln\frac{2+4e+e-2}{e-2} - \ln\frac{1+2e-2(e-2)}{e-2} =$$
$$\ln\frac{5e}{e-2} - \ln\frac{5}{e-2} =$$
$$\ln\frac{\frac{5e}{e-2}}{\frac{5}{e-2}} = \ln\frac{5e}{5} = 1.$$

EX 4. For what value of x is $\log(x+5) = \log x + \log 5$?

Solution ☞ If we use the first law of logarithm on the right hand side of the equation, we get $\log(x+5) = \log 5x \Rightarrow x + 5 = 5x \Rightarrow 4x = 5 \Rightarrow x = \frac{5}{4}$.

EX 5. Solve the exponential equation $3^{2x-1} = 17$.

Solution ☞ If we apply the natural logarithm to both sides, we get

$$\ln 3^{2x-1} = \ln 17 \Rightarrow (2x-1)\ln 3 = \ln 17$$
$$\Rightarrow 2x\ln 3 - \ln 3 = \ln 17$$
$$\Rightarrow 2x\ln 3 = \ln 17 + \ln 3$$
$$\Rightarrow x = \frac{\ln 17 + \ln 3}{2\ln 3}.$$

EX 6. Solve the exponential equation $2^{1-3x} = 5^x$.

Solution ☞ If we apply the natural logarithm to both sides, we get

$$\ln 2^{1-3x} = \ln 5^x \Rightarrow (1 - 3x)\ln 2 = x\ln 5$$
$$\Rightarrow \ln 2 - 3x\ln 2 = x\ln 5$$
$$\Rightarrow \ln 2 = x\ln 5 + 3x\ln 2$$
$$\Rightarrow \ln 2 = (\ln 5 + 3\ln 2)x$$
$$\Rightarrow x = \frac{\ln 2}{\ln 5 + 3\ln 2}.$$

EX 7. Solve the exponential equation $2e^{4x-3} - 5 = 71$.

Solution ☞ To isolate the exponential term on one side of the equation, we add 5 to both sides of the equation then we divide by 2 obtaining $e^{4x-3} = 38$. Applying the natural logarithm on both sides, we get

$$\ln e^{4x-3} = \ln 38 \Rightarrow (4x - 3)\ln e = \ln 38$$
$$\Rightarrow 4x - 3 = \ln 38$$
$$\Rightarrow x = \frac{3 + \ln 38}{4}.$$

EX 8. Solve the exponential equation $e^{2x} - 3e^x - 10 = 0$.

Solution ☞ We can write this equation as $(e^x)^2 - 3e^x - 10 = 0$. That is a quadratic equation in e^x. We can factor this equation into $(e^x - 5)(e^x + 3) = 0$.

Therefore,
$e^x - 5 = 0$ or $e^x + 3 = 0 \Rightarrow e^x = 5$ or $e^x = -3$.
The equation $e^x = 5 \Rightarrow x = \ln 5$. But the equation $e^x = -3$ has no solution because $e^x > 0$ for all x. The only solution to the equation is $x = \ln 5$.

EX 9. Solve the exponential equation $9^x - 3^{x+1} = 28$.

> **Solution** ☞ Since $9^x = 3^{2x} = (3^x)^2$ and $3^{x+1} = 3 \cdot 3^x$, we can write this equation as $(3^x)^2 - 3 \cdot 3^x - 28 = 0$. Again we have a quadratic equation in 3^x. We can factor this equation into $(3^x - 7)(3^x + 4) = 0$.
>
> Therefore,
> $3^x - 7 = 0$ or $3^x + 4 = 0 \Rightarrow 3^x = 7$ or $3^x = -4$.
> The equation $3^x = 7 \Rightarrow x = \log_3 7 = \dfrac{\ln 7}{\ln 3}$. But the equation $3^x = -4$ has no solution because $3^x > 0$ for all x. The only solution to the equation is $x = \dfrac{\ln 7}{\ln 3}$.

EX 10. Solve the equation $5xe^x - x^2 e^x = 0$.

> **Solution** ☞ If we factor e^x, we get $(5x - x^2)e^x = 0$. Since $e^x > 0$ for all x, $e^x \neq 0$. Therefore, $5x - x^2 = 0 \Rightarrow x(5 - x) = 0 \Rightarrow x = 0$ or $x = 5$.

EX 11. ✷ (F02#19) Find the solution(s) (if any exist) of the equation $\log_2(x+1) = \log_2(x) + 3$.

> **Solution** ☞ $\log_2(x + 1) = \log_2(x) + 3 \Rightarrow \log_2(x + 1) - \log_2(x) = 3 \Rightarrow \log_2 \dfrac{x+1}{x} = 3 \Rightarrow \dfrac{x+1}{x} = 2^3 \Rightarrow x + 1 = 8x \Rightarrow 1 = 7x \Rightarrow x = 1/7$.
>
> ✓ Checking this, we see that both $\log_2(1/7+1) = \log_2(8/7)$ and $\log_2(1/7) + 3 = \log_2(1/7) + \log_2(8) = \log_2(1/7 \cdot 8) = \log_2(8/7)$. Thus, $x = 1/7$ is the solution.

Name:_____

Section Number:_____

Date:_____

Chapter 25 Exercises

Solve the logarithmic equations on numbers 1 - 7.

1. $\log_3(4x - 1) = 2$.

2. $\log(x + 7) = 3$.

3. $\log_2(6x + 15) - \log_2 x = 4$.

4. $\ln(2x - 5) = 3$.

5. $\log_4(x + 5) + \log_4(x - 1) = 2.$

6. $\ln 2x - \ln(x + 1) = -1.$

7. $\log_3(2x - 1) + \log_3(x - 4) = 2.$

Solve the exponential equations on numbers 8 - 12.

8. $3^{1-2x} = 7.$

9. $2e^{5x-1} - 3 = 9$.

10. $\left(1 + \dfrac{0.075}{12}\right)^{12t} = 3$.

11. $e^{2x} - 4e^x + 3 = 0$.

12. $2^{3x+1} = 5^x$.

Express y as a function of x for numbers 13 - 14. The constant c is a positive number.

13. $\ln y = \ln x + \ln(x - 1) + c$.

14. $\ln y = 2x + c$.

For numbers 15 - 21, show all work and print the answer in the space provided.

15. The solution to the equation $\log(2x - 1) = 4$.

A.) $x = 41$ B.) $x = 65/2$ C.) $x = 13/2$

D.) $x = 40$ E.) None of these.

Answer:_____

16. The solution to the equation $2^{x+1} = 5$

A.) $\dfrac{\ln 3}{\ln 2}$
B.) $\dfrac{\ln 5}{\ln 2} - 1$
C.) $\dfrac{\ln 2}{\ln 5} + 1$

D.) $1 - \ln\left(\dfrac{5}{2}\right)$
E.) None of these.

Answer:_____

17. Solve $\ln\sqrt{x+5} = 2$ for x.

A.) $x = 2^e - 5$
B.) $x = e^2 - 5$
C.) $x = e^4 - 5$

D.) $x = (e^2 - 5)^2$
E.) None of these.

Answer:_____

18. Solve the exponential equation $e^{2x-1} = 5$ for x.

A.) $\dfrac{2 - \ln 5}{2}$
B.) $e^{1/2} - 5$
C.) $1 + \log_3 5$

D.) $\dfrac{1 + \ln 5}{2}$
E.) None of these.

Answer:_____

19. Solve the logarithmic equation $\log_2(\ln x)) = -1$ for x.

A.) $\sqrt{x}$ B.) e^{-2} C.) $\dfrac{1}{e}$

D.) $\sqrt{e}$ E.) None of these.

Answer:_____

20. Solve the logarithmic equation $\log(x - 1) + \log(x + 2) = 1$ for x.

A.) $x = -7/3$ B.) $x = 5.5$ C.) $x = -3$

D.) $x = -4, 3$ E.) $x = 3$

Answer:_____

21. Solve the logarithmic equation $\log_\pi(x - 2) + \log_\pi x = \log_\pi(x + 4)$ for x.

A.) $x = \pi$ B.) $x = \pi^2$ C.) $x = 1$

D.) $x = 4$ E.) None of these.

Answer:_____

Chapter 26

Applied Problems with Exponential and Logarithmic Functions

26.1 Notes about Applied Problems with Exponential and Logarithmic Functions

In this chapter we will learn how to use exponential and logarithmic functions to model exponential growth and decay. This is very useful in application to financial investments, carbon dating, population growth, decay of radioactive materials and many other phenomena. Examples of other types are also shown.

Compound Interest If a principal P is invested at an annual interest rate r for a period of t years, then the amount A of the investment is given by

$A = P\left(1 + \dfrac{r}{n}\right)^{nt}$ if the interest is compounded n times per year and

$A = Pe^{rt}$ if the interest is compounded continuously.

Exponential Growth To model the population growth of a city of the growth of bacterium or any other similar phenomena we use the exponential growth model $N = N_0 e^{rt}$, where N is the population at a time t, N_0 is the initial population or the population at time t_0 and r is the relative rate of growth.

Radioactive Decay The rate of decay of radioactive substances varies directly as the mass of the substance. This is similar to exponential growth, except the mass of radioactive material decreases. The amount A of a radioactive substance remaining after t years is given by $A = A_0 e^{rt}$, where A_0 is the initial amount present and r is the annual rate of decay of that substance (where r is a negative number). The *half-life* of a radioactive substance is the amount of time it takes for one-half of the substance to decay.

26.2 ❖ Errors to Avoid

Don't forget to convert percentages to decimals.

26.3 Examples

EX 1. If a bank pays 3.2% annual interest compounded monthly then what will a deposit of $2,500 be worth after 5 years?

Solution ☞ We use the formula $A = P\left(1+\dfrac{r}{n}\right)^{nt}$. We are given $r = 3.2\% = 0.032$, $P = \$2,500$, $t = 5$ years and $n = 12$ since the interest is compounded 12 times per year. Thus, $A = 2,500\left(1 + \dfrac{0.032}{12}\right)^{12(5)} = 2,933.15$.

EX 2. If $10,000 is invested in a bank account paying 4% compounded continuously. What amount will be in the account at the end of 9 years?

Solution ☞ We have to use the $A = Pe^{rt}$ with $P = 10,000$, $r = 0.04$ and $t = 9$. Thus, $A = 10,000e^{(0.04)(9)} = 14333.29$.

EX 3. Find the time required for an investment of $8,000 to grow to $12,000 at an annual interest rate of 6.75% per year compounded continuously.

Solution ☞ We have to use the formula $A = Pe^{rt}$ with $A = 12,000$, $P = 8,000$ and $r = 0.0675$ to find t.

$$12,000 = 8,000e^{0.0675t} \Rightarrow e^{0.0675t} = \frac{120000}{8000}$$
$$\Rightarrow \ln e^{0.0675t} = \ln 1.5$$
$$\Rightarrow 0.0675t \ln e = \ln 1.5 \qquad\qquad (\ln e = 1)$$
$$\Rightarrow t = \frac{\ln 1.5}{0.0675} \Rightarrow t \approx 6 \text{ years.}$$

EX 4. The population of Denton, Texas is growing according to the function $N = 73,225e^{0.033t}$, where t is the number of years since $2,000$.

   (a) What will the population be in 2005?

   Solution ☞ In the year 2005, $t = 5$. Therefore, $N = 73,225e^{0.033(5)} = 86,361$.

   (b) In what year will the population of Denton reach 200,000?

   Solution ☞ If we replace N with 200,000, we get the exponential equation $200,000 = 73,225e^{0.033t}$. To solve for t, we divide both sides by 73,225 yielding

$$\frac{200,000}{73,225} = e^{0.033t} \Rightarrow \ln\left(\frac{200,000}{73,225}\right) = 0.033t$$

$$\Rightarrow t = \frac{\ln\frac{200,000}{73,225}}{0.033} \approx 30.45 \text{ years}.$$

   This is between the years 2030 and 2031.

EX 5. A culture contains 100 bacteria initially and doubles every 90 minutes.

   (a) Find a function that models the number of bacteria N after t minutes.

   Solution ☞ Using the exponential growth model $N = N_0 e^{rt}$ with $N_0 = 100$, we can say that $N = 100e^{rt}$. We need to find r. We know when $t = 90$ minutes, the culture is double the initial size. Therefore, $N = 200$ when $t = 90$. Replacing N with 200 and t with 90 in the model, we get
$200 = 100e^{r \cdot 90} \Rightarrow e^{90r} = \frac{200}{100} \Rightarrow e^{90r} = 2 \Rightarrow \ln e^{90r} = \ln 2 \Rightarrow 90r$
$= \ln 2 \Rightarrow r = \frac{\ln 2}{90} \approx 0.0077$.

   Therefore, $N = 100e^{\frac{\ln 2}{90}t}$ with t in minutes.

   (b) Find the number of bacteria after 24 hours.

   Solution ☞ After 24 hours, $t = 24 \cdot 60 = 1,440$ and $N = 100e^{\frac{\ln 2}{90}(1440)} = 6,553,600$

   (c) After how many hours will the number of bacteria be 100,000,000?

   Solution ☞ If $N = 100,000,000$, then $100,000,000 = 100e^{\frac{\ln 2}{90}t} \Rightarrow$
$e^{\frac{\ln 2}{90}t} = 1,000,000 \Rightarrow \frac{\ln 2}{90}t = \ln(1,000,000) \Rightarrow t = \frac{\ln(1,000,000)}{\frac{\ln 2}{90}} \approx$
$1,793.84 \text{ minutes}$.

   Thus, the number of bacteria will be 100,000,000 after about 30 hours.

EX 6. ★ (S04#17) The rate at which rumors spread around Big U. is given by $N = \dfrac{25000}{1 + 100e^{-0.05t}}$, where N is the number of students that have heard the rumor and t is the time in days.

(a) How many students at Big U. have heard the rumor after 10 days? Round to the nearest student.

Solution ☞ Here $t = 10$. So $N = \dfrac{25000}{1 + 100e^{-0.05(10)}} \approx 405.49$ or 405 students.

(b) How many days will it be before 18,000 students at Big U. have heard the rumor? Round to the nearest day.

Solution ☞ Here we want to find t if $N = 18,000$. Substituting in $N = 18,000$ and solving gives

$$18000 = \frac{25000}{1 + 100e^{-0.05(t)}} \Rightarrow$$
$$18000(1 + 100e^{-0.05t}) = 25000 \Rightarrow$$
$$1 + 100e^{-0.05t} = \frac{25000}{18000} \Rightarrow$$
$$100e^{-0.05t} = \frac{25000}{18000} - 1 \Rightarrow$$
$$e^{-0.05t} = \frac{\frac{25000}{18000} - 1}{100} \Rightarrow$$
$$-0.05t = \ln\left(\frac{\frac{25000}{18000} - 1}{100}\right) \Rightarrow$$
$$t = \frac{\ln\left(\frac{\frac{25000}{18000} - 1}{100}\right)}{-0.05} \approx 110.99 \text{ or } 111 \text{ days.}$$

EX 7. The mass remaining after t years from a 50-gram sample of a certain radioactive substance is given by $m = 50e^{-0.0015t}$.

(a) How many grams will remain after 100 years?

Solution ☞ When $t = 100$, $m = 50e^{-0.0015(100)} \approx 43$ grams.

(b) How long will it take for the sample to decay to a mass of 10 grams?

Solution ☞ $m = 10 \Rightarrow 10 = 50e^{-0.0015t} \Rightarrow e^{-0.0015t} = \frac{10}{50} \Rightarrow$ $\ln e^{-0.0015t} = \ln \frac{1}{5} \Rightarrow -0.0015t = \ln \frac{1}{5} \Rightarrow t = \frac{\ln \frac{1}{5}}{-0.0015} \approx 1,073$ years.

(c) What is the half-life of this substance?

Solution ☞ The half-life of a substance is the time it takes for half
the mass to decay. Here the original mass is 50 grams. To find the
half-life we need to replace N with 25 (half of 50) and solve for t.

$$25 = 50e^{-0.0015t} \Rightarrow e^{-0.0015t} = \frac{25}{50}$$
$$\Rightarrow \ln e^{-0.0015} = \ln(1/2)$$
$$\Rightarrow -0.0015t = \ln(1/2)$$
$$\Rightarrow t = \frac{(1/2)}{-0.0015} \approx 462 \text{ years.}$$

So the half-life of this substance is about 462 years.

EX 8. An ancient wooden statue contains 73% of the original amount of carbon-14.
If the half-life of carbon-14 is 5,730 years, how old is the statue?

Solution ☞ The amount of carbon-14 remaining in the statue after t years
can be modeled by the function $m = m_0 c^{-rt}$, where m_0 is the initial amount
of carbon-14, r is the rate of decay and t is the number of years.

We can use the fact that the half-life of carbon-14 is 5,730 to find r. That is
$m = \frac{m_0}{2}$ where $t = 5,730 \therefore \frac{m_0}{2} = m_0 e^{-r5730} \Rightarrow \frac{1}{2} = e^{-5730r} \Rightarrow \ln e^{-5730r} = $
$\ln(1/2) \Rightarrow -5730r = -\ln 2 \Rightarrow r = \frac{\ln 2}{5730}$.
Therefore, $m = m_0 e^{-\frac{\ln 2}{5730}t}$.

Now since the statue contains 73% of the original amount of carbon-14,
$0.73m_0 = m_0 e^{-\frac{\ln 2}{5730}t} \Rightarrow \frac{0.73m_0}{m_0} = e^{-\frac{\ln 2}{5730}t} \Rightarrow \ln 0.73 = \ln e^{-\frac{\ln 2}{5730}t} \Rightarrow -\frac{\ln 2}{5730}t = $
$\ln 0.73 \Rightarrow t = \frac{\ln 0.73}{-\frac{\ln 2}{5730}} \approx 2601.6$.

Therefore the statue is about 2,602 years old.

Chapter 26 Exercises

Answer questions 1 - 5 in the space provided. Be sure to show all work.

1. Ryan invests $8,000 in a bank that pays 4.65% interest per year, compounded monthly.

   (a) Find the amount in the account after 5 years.

   (b) How long will it take for the amount to be $12,000?

2. Sara invests $10,000 in an account that pays 5% interest per year compounded continuously.

   (a) Find the amount in the account after 10 years.

   (b) How long will it take for the amount to be $21,170?

   (c) How long will it take for the original amount to quadruple?

3. The number of bacteria in a culture can be modeled by $N = 300e^{0.15t}$, where t is measured in hours.

    (a) Find the number of bacteria at time $t = 0$.

    (b) Find the number of bacteria after 5 hours.

    (c) After how many hours will the number of bacteria reach $6,000$?

4. The population N of a small city can be approximated by the function $N = 2,500e^{0.02t}$, where t is the number of years after the year 2000.

    (a) Find the population of the city in the year 2000.

    (b) Find the population of the city in the year 2010.

    (c) In what year the population of the city will reach $4,120$?

5. A wooden arrow found in an old cave contains about 81% of the original amount of carbon-14. How old is the arrow?

For numbers 6 - 10, show all work and print the answer in the space provided.

6. Lisa invests $20,000 at an annual interest rate of 7.5% compounded monthly, how much will she have in 15 years.

    A.) $21,959.3   B.) $61,604.34   C.) $59,177.55
    D.) $61,389.03   E.) None of these.

                                    Answer:_____

7. How many years will it take for an investment of $2,000$ to grow to $5,000$ if the interest is 6.5% compounded continuously (round your answer to the nearest year)?

   A.) 10 years   B.) 12 years   C.) 14 years

   D.) 16 years   E.) None of these.

Answer:_____

8. The number N of students of a small university can be approximated by the function $N = 1250e^{0.04t}$, where t is the number of years after the year 1980. In what year will the number of students reach 4,150?

   A.) 2000   B.) 2004   C.) 2008

   D.) 2010   E.) The number of students will never reach 4,150.

Answer:_____

9. The number of bacteria N in a culture can be modeled by the function $N = 80e^{0.2t}$, where t is measured in hours. About how long will it take for the culture to reach 16,350?

A.) 20 hours and 35 minutes   B.) 26 hours and 36 minutes

C.) 30 hours and 37 minutes   D.) 33 hours and 38 minutes

E.) None of these.

Answer:_____

10. In 1985 about 20 jars made out of clay were found during construction of a highway in Kalamoun, Lebanon. One of the jars had cotton in it that contained 54% of the original carbon-14. Estimate the age of the cotton (round your answer to the nearest year).

A.) $5,094$ years   B.) $5,050$ years   C.) $5,000$ years

D.) $4,500$ years   E.) None of these.

Answer:_____

Chapter 27

Systems of Equations

27.1 Notes about Systems of Equations

Definition: A *system of equations* is a set of equations with common variables. Most of the systems we will consider consist of 2 equations in two variables.

Definition: To *solve* a system of equations is to find the values of all the variables that make the system true.

Definition: A system of equations is said to be *consistent* if it has at least one solution. A system is said to be *inconsistent* if it has no solution.

27.1.1 Substitution Method:

1. Pick one equation and solve for one of the variables in terms of the other.

2. Substitute the expression found in Step 1 into the equation obtaining an equation in one variable. Solve the equation.

3. Back-substitute the value found in Step 2 into the expression obtained in Step 1 to find the other variable.

27.1.2 Elimination Method:

To use the elimination method we may use the following steps:

1. Pick the variable you want to eliminate and then multiply one or both equations by suitable numbers so that the coefficient of the variable you want to eliminate in one equation is the opposite of that coefficient in the other equation.

2. Add the 2 equations to eliminate one of the variables and solve the resulting equation for the remaining variable.

3. Back-substitute the value obtained in step 2 into one of the original equations and solve for the other variable.

27.2 Examples

EX 1. Solve the system of equations

$$\begin{cases} 2x - y = 0 \\ x^2 + y^2 = 5 \end{cases}$$

Solution ☞ First we solve the first equation for y in terms of x.

$2x - y = 0 \Rightarrow y = 2x$.

Next we substitute $2x$ for y in the second equation

$$x^2 + (2x)^2 = 5 \Rightarrow$$
$$x^2 + 4x2 = 5 \Rightarrow$$
$$5x^2 = 5 \Rightarrow$$
$$x^2 = 1 \Rightarrow x = \pm 1.$$

We get 2 values for x, -1 and 1. To find the values of y, we will use back substitution by plugging the x-values into the equation $y = 2x$.

When $x = -1$, $y = 2(-1) = -2 \Rightarrow (-1, -2)$ is a solution.
When $x = 1$, $y = 2(1) = 2 \Rightarrow (1, 2)$ is a solution.

✎ To check our solution, we substitute $(-1, -2)$ into both equations giving:

$$\begin{cases} 2(-1) - (-2) = -2 + 2 = 0 \\ (-1)^2 + (-2)^2 = 1 + 4 = 5 \end{cases}$$

We see that the solution $(-1, -2)$ satisfies both equations.

Checking $(1, 2)$ into both equations giving:

$$\begin{cases} 2(1) - (2) = 2 - 2 = 0 \\ (1)^2 + (2)^2 = 1 + 4 = 5 \end{cases}$$

We see that the solution $(1, 2)$ satisfies both equations and therefore both $(-1, -2)$ and $(1, 2)$ are solutions.

Note that the solutions $(-1, -2)$ and $(1, 2)$ are the points of intersection of the two equations of the system. Because of this, it is also possible to check your answers by graphing both equations and determining the points of intersection.

EX 2. Solve the system of equations

$$\begin{cases} 7x + 2y - -4 \\ 2x + y = 1 \end{cases}$$

Solution ☞ Solving for y in terms of x in the second equation, we get $y = 1 - 2x$.

Next we replace y with $1 - 2x$ in the first equation obtaining $7x + 2(1 - 2x) = -4$. Continuing solving for x yields $7x + 2(1 - 2x) = -4 \Rightarrow 7x + 2 - 4x = -4 \Rightarrow 3x + 2 = -4 \Rightarrow 3x = -6 \Rightarrow x = -2$.

Back substituting to solve for y gives $y = 1 - 2(-2) = 5$.

Thus, the solution is $(-2, 5)$.

✎ We check our solution by replacing x with -2 and y with 5 in the system.

$$\begin{cases} 7(-2) + 2(5) = -14 + 10 = -4 \\ 2(-2) + (5) = -4 + 5 = 1 \end{cases}$$

We see that $(-2, 5)$ satisfies both equations and therefore is the solution.

EX 3. Solve the system

$$\begin{cases} x^2 - y = 1 \\ x - y = -1 \end{cases}$$

Solution ☞ We begin by solving for y in terms of x in the second equation giving $x - y = -1 \Rightarrow x + 1 = y$ or $y = x + 1$.

Next replace y with $x + 1$ in the first equation $x^2 - y = 1 \Rightarrow x^2 - (x + 1) = 1 \Rightarrow x^2 - x - 2 = 0$. This is a quadratic equation that can be solved by factoring. The solution is below.

$x^2 - x - 2 = 0 \Rightarrow (x - 2)(x + 1) = 0 \Rightarrow x - 2 = 0$ or $x + 1 = 0 \Rightarrow x = 2$ or $x = -1$.

To find y back substitute $x = 2$ and $x = -1$ to the equation $y = x + 1$.

$x = 2 \Rightarrow y = (2) + 1 = 3 \Rightarrow (2, 3)$ is a solution.
$x = -1 \Rightarrow y = (-1) + 1 = 0 \Rightarrow (-1, 0)$ is a solution.

Checking both solutions shows that each works.

EX 4. ✯ (S03#6) Find $3x - y$ if (x, y) is the solution to the system of linear equations.

$$\begin{cases} 3x - 2y = 3 \\ x + 4y = 8 \end{cases}$$

A.) 8   B.) 4.5   C.) 3   D.) 1   E.) 2

Solution ☞ To solve this problem, we will first find the values of x and y that solve the system and then find the value of $3x - y$.

We begin by solving the second equation for x in terms of y giving $x + 4y = 8 \Rightarrow x = 8 - 4y$.

Next replace x with $8 - 4y$ in the first equation giving $3(8 - 4y) - 2y = 3$. Solving this equation for y gives $3(8 - 4y) - 2y = 3 \Rightarrow 24 - 12y - 2y = 3 \Rightarrow -14y = -21 \Rightarrow y = 1.5$.

Back substituting to find x gives $x = 8 - 4(1.5) = 8 - 6 = 2$. Next, the x- and y-values of the solution $(2, 1.5)$ need to be substituted into $3x - y$ yielding $3(2) - 1.5 = 6 - 1.5 = 4.5$.

EX 5. Solve the system

$$\begin{cases} 2x + y = 5 \\ 3x^2 + y = 1 \end{cases}$$

Solution ☞ We can find y in terms of x in the first equation giving $2x + y = 5 \Rightarrow y = 5 - 2x$.

Replacing y with $5 - 2x$ in the second equation gives $3x^2 + y = 1 \Rightarrow 3x^2 + (5 - 2x) = 1 \Rightarrow 3x^2 - 2x + 4 = 0$ that is a quadratic equation in x. Using the quadratic equation to solve this gives

$$x = \frac{-b \pm \sqrt{b^2 - 4ac}}{2a} = \frac{--2 \pm \sqrt{(-2)^2 - 4(3)(4)}}{2(3)}$$
$$= \frac{2 \pm \sqrt{4 - 48}}{6} = \frac{2 \pm \sqrt{-44}}{6}.$$

This equation does not have a real solution. Therefore the system has no solution. In this case, we say the system is inconsistent. In this case the graphs of the two equations of the systems do not meet.

EX 6. Solve the system

$$\begin{cases} x - 3y = 2 \\ -5x + 15y = -10 \end{cases}$$

Solution From the first equation, $x - 3y = 2$, we can find x in terms of y. $x - 3y = 2 \Rightarrow x = 2 + 3y$.

Replacing x with $2 + 3y$ in the second equation and solving for y yields $-5(2 + 3y) + 15y = -10 \Rightarrow -10 - 15y + 1y = -10 \Rightarrow -10 = -10$, which is always true. Therefore, y can be any real number and x will always be equal to $2 + 3y$. The solution is the set of all ordered pairs of the form $(2 + 3y, y)$, where y is any real number. Hence, the system has infinitely many solutions. In this case, we say the equations are dependent.

Looking back at the original system, you may notice that the second equation is the first equation multiplied by -5. Because of this the graphs of the equations of the system coincide.

EX 7. Use the elimination method to solve the system

$$\begin{cases} 2x + y & = 1 \qquad\qquad (1) \\ 3x + 2y & = 4 \qquad\qquad (2) \end{cases}$$

Solution ☞ If we choose to eliminate the y-term, we can multiply the first equation by -2 yielding the following set of equations:

$$\begin{cases} -4x - 2y = - 2 \qquad\qquad (1) \\ 3x + 2y = 4 \qquad\qquad (2) \end{cases}$$

Next we add the two equations and solve for x giving $-x = 2 \Rightarrow x = -2$. Now we back-substitute $x = -2$ into one of the original equations to find y, choosing the first one gives $2x + y = 1 \Rightarrow 2(-2) + y = 1 \Rightarrow -4 + y = 1 \Rightarrow y = 1 + 4 \Rightarrow y = 5$. Thus, $(-2, 5)$ is the solution to the system.

✏ To verify this, we substitute the x- and y-values $(-2, 5)$ into the original equations.

$$\begin{cases} 2(-2) + (5) = -4 + 5 = 1 \qquad\qquad (1) \\ 3(-2) + 2(5) = -6 + 10 = 4 \qquad\qquad (2) \end{cases}$$

We see that both equations are true and therefore the solution is $(-2, 5)$.

EX 8. Use the elimination method to solve the system

$$\begin{cases} x + y & = 4 \\ x^2 - 2x - y & = 2 \end{cases}$$

Solution ☞ We can eliminate the y-term in here by just adding the two equations of the system. If we do so, we have $x^2 - x = 6 \Rightarrow x^2 - x - 6 = 0 \Rightarrow (x + 2)(x - 3) = 0 \Rightarrow x = -2$ or $x = 3$. Next we use the first equation $x + y = 4$ to find the y-values.

If $x = -2$, then $(-2) + y = 4 \Rightarrow y = 6$.
If $x = 3$, then $(3) + y = 4 \Rightarrow y = 1$.

$\therefore$ $(-2, 6)$ and $(3, 1)$ are the solutions of the system.

EX 9. Use the elimination method to solve the system

$$\begin{cases} 5x - 3y & = 7 \\ 4x + 2y & = -1 \end{cases}$$

Solution ☞ If we choose to eliminate the y-term, then we can multiply the first equation by 2 and the second equation by 3 to obtain coefficients for y that differ only in sign.

$$\begin{cases} 10x - 6y & = 14 \\ 12x + 6y & = -3 \end{cases}$$

Next we add the two equations and solve for x yielding $22x = 11 \Rightarrow x = \dfrac{1}{2}$.

Now to find the value of y, we can replace x with $\dfrac{1}{2}$ in one of the original two equations. Choosing the second equation gives $4x + 2y = -1 \Rightarrow 4\left(\dfrac{1}{2}\right) + 2y = -1 \Rightarrow 2y = -3 \Rightarrow y = \dfrac{3}{2}$.

Therefore, $\left(\dfrac{1}{2}, \dfrac{-3}{2}\right)$ is the solution of the system.

Chapter 27 Exercises

For numbers 1 - 7, solve the given systems of equations.

1.

$$\begin{cases} 2x + y = 1 \\ 3x - y = 9 \end{cases}$$

2.

$$\begin{cases} 2x - y = -1 \\ x^2 + y = 4 \end{cases}$$

3.

$$\begin{cases} 5x + y = 1 \\ 4x + 6y = -7 \end{cases}$$

4.

$$\begin{cases} 2x - y = -5 \\ x^2 + y^2 = 2 \end{cases}$$

5.

$$\begin{cases} x = 2y - 1 \\ 3x - 6y = 7 \end{cases}$$

6.

$$\begin{cases} 2x - 3y = 1 \\ 6x = 9y + 3 \end{cases}$$

7. ★ (S03#18)

$$\begin{cases} y + x^2 = 4x \\ y + 4x = 16 \end{cases}$$

For numbers 8 - 11, show all work and print the answer in the space provided.

8. The solution of the system of equations

$$\begin{cases} \frac{1}{2}x + 3y = -5 \\ x - y = 11 \end{cases}$$

A.) $x = 2, y = -2$ B.) $x = 12, y = 1$
C.) $x = -3, y = 8$ D.) $x = 8, y = -3$
E.) The system has no solution.

Answer:_____

9. Find all the solutions to the system

$$\begin{cases} 2x + y = -1 \\ x^2 - y = 2 \end{cases}$$

A.) $(2, -5)$ and $(3, 7)$ B.) $(0, -2)$ and $(5, -11)$
C.) $(1, -1)$ and $(-3, 7)$ D.) $(\frac{1}{2}, -2)$ and $(\frac{1}{2}, \frac{1}{4})$
E.) None of these.

Answer:_____

10. Find $2a + b$ if (a, b) is a solution to the system

$$\begin{cases} 2a - 3b = 17 \\ a + 2b = -2 \end{cases}$$

A.) 1   B.) -3 C.) -2 D.) 5   E.) 4

Answer:_____

11. Find $a - 2b$ if (a, b) is a solution to the system

$$\begin{cases} 3a + 4b = 7 \\ \dfrac{1}{2}a - 3b = 3 \end{cases}$$

A.) $\dfrac{-1}{2}$ B.) 3   C.) -1 D.) $\dfrac{5}{2}$ E.) 4

Answer:_____

283

Chapter 28

Systems of Inequalities

28.1 Notes about Systems of Inequalities

In Chapter 7 we learned how to solve linear inequalities in one variable. Unlike linear inequalities in one variable, the solution set of a system linear inequality in two variables is a collection of ordered pairs or points in the x, y-plane. To represent this solution set, we are going to graph an area of the x, y-plane. t

To graph a solution set of an inequality in two variables, follow the steps below:

1. Replace the inequality signs with equal signs and sketch the graph of the resulting equation. With strict inequalities, used dashed lines rather than solid lines. Otherwise, use a solid line.

2. For each inequality, choose a *test point*. The test point must be a point in the plane that is not on the line associated with the inequality.

3. Determine whether each test point satisfies the original inequality. If the test point makes the inequality true, then shade the region to denote that every point of this region is in the solution set. If the coordinates of the test point do not satisfy the inequality, then we shade the other region, which is the solution set.

4. Shade the area of the x, y-plane for which all of the arrows point in the same direction. This shaded area is the *solution of the system*.

28.2   Examples

EX 1. Graph the solution set of the inequality $2x + y \leq 1$.

Solution ☞ First we replace the $\leq$ with an $=$ sign obtaining the linear equation $2x + y = 1$. Next we sketch the graph of $2x + y = 1$ or $y = -2x + 1$. The graph is a line with slope -2 and y-intercept $(0, 1)$. We draw a solid line since the inequality ($\leq$) is not strict. This line represents the plan into two regions. One of the regions is the solution to the inequality. To determine which region is the solution to the inequality. To determine which region is the solution we pick a test point, say $(-1, 1)$ replace x with -1 and y with 1 in the original inequality $2x + y \leq 1$. Is $2(-1) + 1 \leq 1$? Well, $2(-1) + 1 = -2 + 1 = -1 \leq 1$ is true. Therefore, $(-1, 1)$ is in the solution set. We shade the region that contains the point $(-1, 1)$. The solid line indicates that the points on the line are part of the solution set.

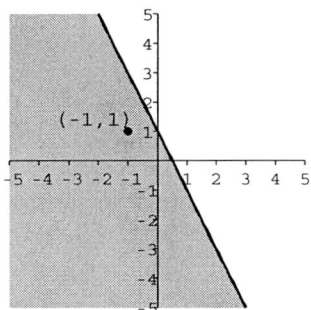

Figure 28.1: Graph of $2x + y \leq 1$.

EX 2. Graph the solution of $y > 1$.

Solution ☞ First, we replace the $>$ with an $=$ sign and graph the line $y = 1$. Let us pick $(1, 2)$ as a test point. If we replace y with 2 (x with 1 would be necessary if there had been an x-term in the original inequality) in the inequality $y > 1$ gives $2 > 1$ which is true so $(1, 2)$ is on the solution side. Thus the half-plane above $y = 1$ represents the solution of the inequality.

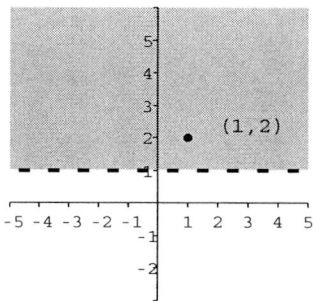

Figure 28.2: Graph of $y > 1$.

EX 3. Graph the inequality $x^2 - y \geq 0$.

Solution ☞ First we graph the equation $x^2 - y = 0$ or $y = x^2$. Since the origin is on the graph, we cannot use the origin as a test point. So we pick the point $(0, 3)$. If we replace x with 0 and y with 3 in $x^2 - y \geq 0$, we get $(0)^2 - (3) \geq 0$ which is not true. Therefore, $(0, 3)$ is not on the solution side and we shade the other side which is the solution set.

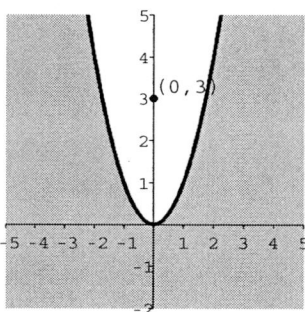

Figure 28.3: Graph of $x^2 - y \geq 0$.

EX 4. Graph the solution set of the system

$$\begin{cases} x + y \geq 1 \\ 2x - y \leq 2 \end{cases}$$

Solution ☞ The solution set of this system is where the solution sets of the two systems overlap. So let us graph the solutions of both inequalities on the same x, y-coordinate system.

- For the first inequality, graph the solid line $x + y = 1$ or $y = -x + 1$. Picking $(0,0)$ as a test point gives $(0) + (0) \geq 1$, which is not true $\therefore$ $(0,0)$ is not on the solution side, so we shade the other side.

- For the second inequality $2x - y = 2$ or $y = 2x - 2$. If we graph the line $y = 2x - 2$ and picking the same test point $(0,0)$ gives $2 \cdot 0 - 0 \leq 2$, which is true. Therefore, $(0,0)$ is on the solution side of the inequality $2x - 4 \leq 2$.

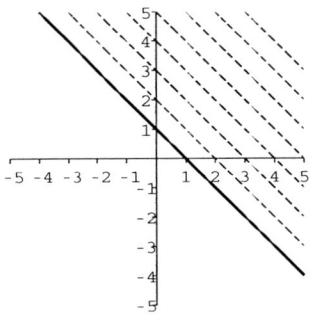
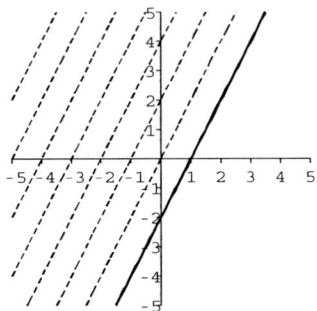

Figure 28.4: Solution set of the inequality $x + y \geq 1$.

Figure 28.5: Solution set of the inequality $2x - y \leq 2$.

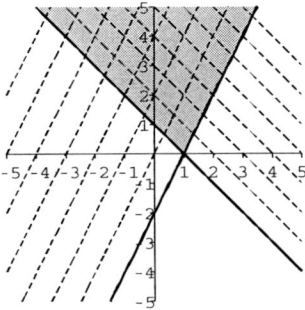

Figure 28.6: Solution set of $x + y \geq 1$ and
$2x - y \leq 2$.

Now we can easily see where the two solution sets intersect that is the systems solution set.

EX 5. Graph the solution set of the system

$$\begin{cases} x \geq 0 \\ y \geq 0 \\ 2x + y \leq 3 \end{cases}$$

Solution ☞ For the first inequality, we graph the vertical line $x = 0$ that is the y-axis. Picking $(1, 1)$ as a test point $1 \geq 0$ true. So the right hand side of the y-axis is the solution of the inequality $x \geq 0$. Similarly the solution to $y \geq 0$ is the set of all points above the x-axis. For the third inequality replace, $\leq$ with an $=$ sign. Using the fact that $2x + y = 3 \Rightarrow y = -2x + 3$. Now we graph the line $y = -2x + 3$ and pick $(0, 0)$ as a test point. Noting that $2 \cdot 0 + 0 \leq 3$ is true implies that $(0, 0)$ is on the solution side of the third inequality. The solution of the system is where all three solution sets of the inequalities overlap. That is the triangle with vertices $(0, 0)$, $(0, 3)$ and $\frac{3}{2}, 0)$.

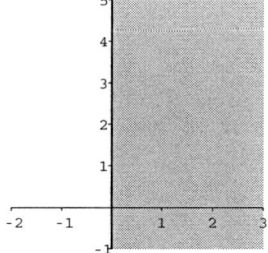

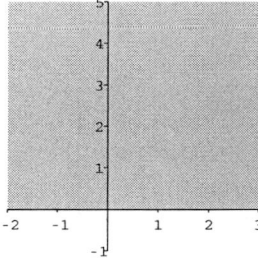

Figure 28.7: Solution set of $x \geq 0$.　　　Figure 28.8: Solution set of $y \geq 0$.

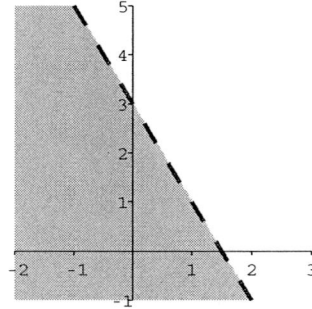

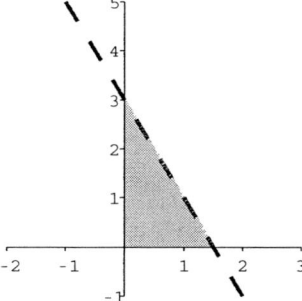

Figure 28.9: Solution set of the in- Figure 28.10: Solution set of $x \geq 0$, equality $2x + y \leq 3$.　　　　　　$y \geq 0$ and $2x + y \leq 3$.

EX 6. Graph the solution set of the system

$$\begin{cases} 2x + y < 0 \\ x^2 - y < 1 \end{cases}$$

Solution ☞ To graph the first inequality, we replace $<$ with an $=$ sign. $2x + y = 0 \Rightarrow y = -2x$. Now graph the dashed line $y = -2x$ and picking $(1, 1)$ as a test point shows $2 \cdot 1 + 1 < 0$ as false. $\therefore$ $(1, 1)$ is not on the solution set side and we shade the other side.

For the second inequality, we graph $x^2 - y = 1$ or $x^2 - 1 = y$, which is the graph of the parabola $y = x^2$ shifted down one unit. Picking $(0, 0)$ as a test point, we get $0^2 - 0 < 1$ which is true. We shade the region above the parabola $y = x^2 - 1$.

Combining both of these graphs together results in the graph below.

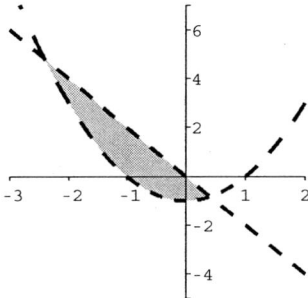

Figure 28.11:

EX 7. ✭ (F03#22) Determine a set of inequalities to describe the shaded region below.

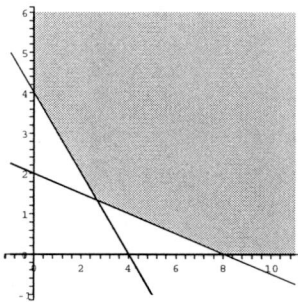

Figure 28.12:

Solution ☞ First notice that while there seems to be two inequalities involved, the graph never leaves quadrant I. Because of this, we know that $x \geq 0$ and $y \geq 0$.

Now dealing with the remaining two inequalities, we note that $(0,4)$ and $(4,0)$ are points on the line forming a boundary of one of the inequalities. Thus, this inequality must be either $y \geq -x+4$ or $y \leq -x+4$. Noting that $(0,0)$ makes the first inequality false, we see that $y \geq -x+4$ represents the inequality passing through $(0,4)$ and $(4,0)$.

Finding the line through $(8,0)$ and $(0,2)$ yields $y = -.25x+2$, which means either $y \geq -.25x+2$ or $y \leq -.25x+2$ is the final inequality. Using $(0,0)$ as a test point makes $y \geq -.25x+2$ false and therefore $y \geq -.25x+2$ is part of the system.

Bringing each of the four inequalities together shows that the system below gives the solution set asked for.

$$\begin{cases} x \geq & 0 \\ y \geq & 0 \\ y \geq -x+4 \\ y \geq -.25x+2 \end{cases}$$

Chapter 28 Exercises

For numbers 1 - 8 graph the solution set on the graph provided.

1. $3x + y \geq 2$.

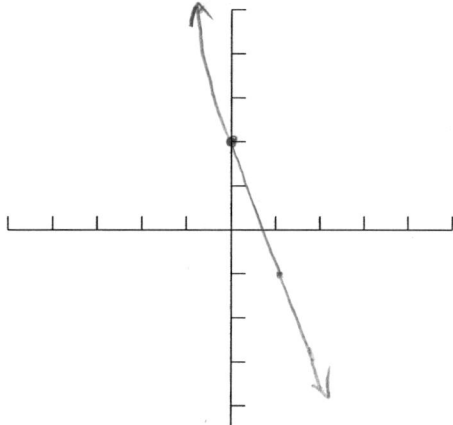

2. $x - 2y < 8$.

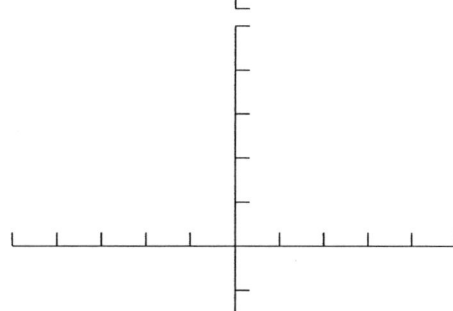

3. $x \leq 3$.

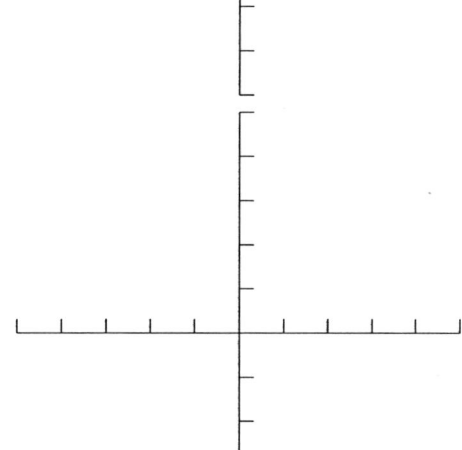

4. $x > 3y$.

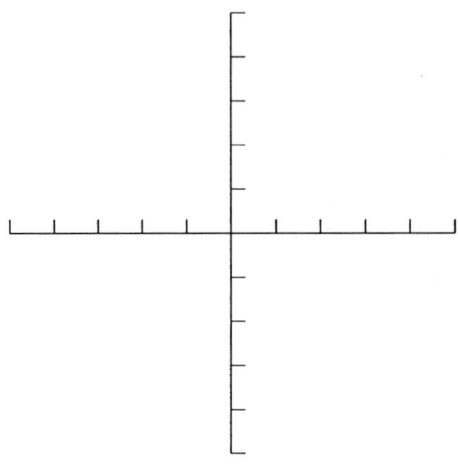

5.

$$\begin{cases} x + y \le 1 \\ x - 2y \ge 4 \end{cases}$$

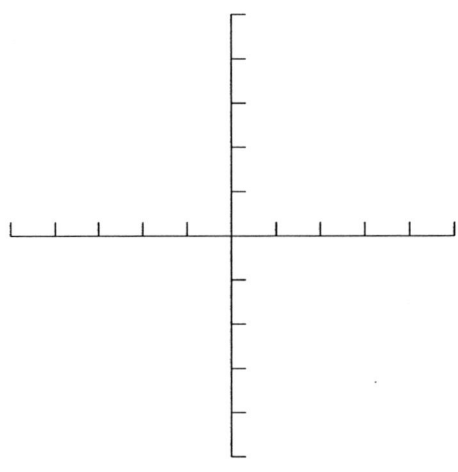

6.

$$\begin{cases} x + y < 3 \\ -3x + 2y < 6 \\ y > 0 \end{cases}$$

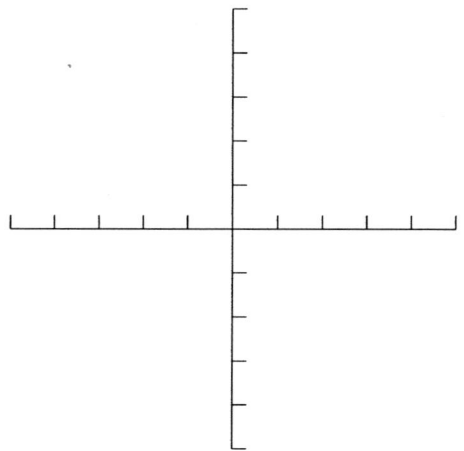

7.

$$\begin{cases} x^2 + y \leq 4 \\ x - 2y \geq 0 \end{cases}$$

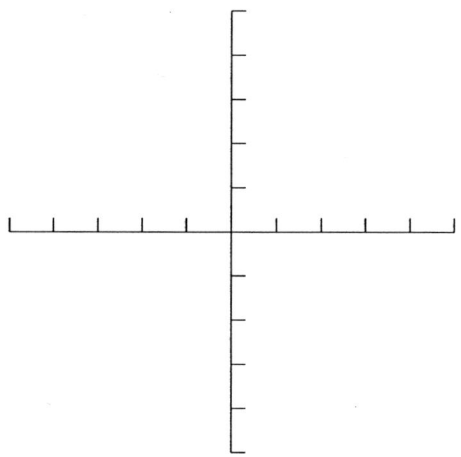

8.

$$\begin{cases} 2x + 3y \le 6 \\ y \ge -1 \\ x > 0 \end{cases}$$

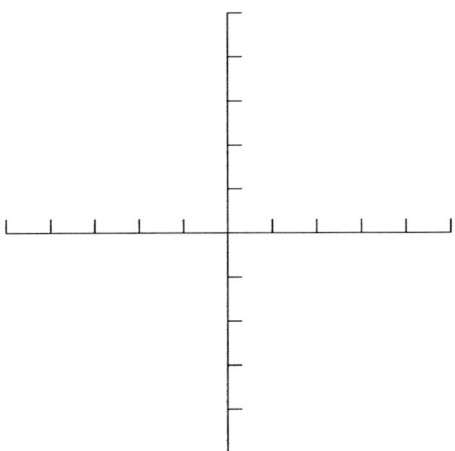

9. Find the system of inequalities for which the solution set is graphed in the Figure 9.

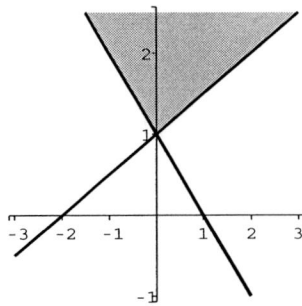

For numbers 10 - 12, show all work and print the answer in the space provided.

10. Which of the following points is not in the solution set of the inequality $5x - 3y < 7$?

   A.) $(-1, 2)$

   B.) $\left(\dfrac{3}{5}, -\dfrac{2}{3}\right)$

   C.) $\left(-\dfrac{2}{5}, -\dfrac{1}{9}\right)$

   D.) $\left(\dfrac{1}{5}, -3\right)$

   E.) $(3, 3)$

Answer:_____

11. The second quadrant is the solution set of which of the following systems of inequalities (note that the second quadrant doesn't include the x or y axes)?

   A.) $\begin{cases} x > 1 \\ y < 1 \end{cases}$

   B.) $\begin{cases} x > 0 \\ y < 0 \end{cases}$

   C.) $\begin{cases} x < 0 \\ y > 0 \end{cases}$

   D.) $\begin{cases} x < 0 \\ y < 0 \end{cases}$

   E.) None of these.

Answer:_____

12. Which of the following points is in the solution set of the system of inequalities below?
$$\begin{cases} x > 0 \\ x - y > -4 \\ 2x + y < 7 \end{cases}$$

   A.) $(-2, 3)$

   B.) $(6, 3)$

   C.) $(1, 5)$

   D.) $(0, 0)$

   E.) $(1, -3)$

Answer:_____

297

Chapter 29

Old Finals

29.1   Fall 2002 Final

1.) Which of the following is the equation of the line through the point $(1,7)$ and parallel to the line passing through the points $(2,5)$ and $(-2,1)$?

A.) $y = x + 7$ B.) $y = 2x + 5$ C.) $y = -2x + 9$

D.) $y = x + 6$ E.) $y = 7x + 1$

2.) Determine whether the function $f(x) = 3x^2 - 18x + 29$ has a maximum or minimum value (if either exists) and find the value of f at the maximum or minimum (if either exists).

A.) minimum; 3   B.) minimum; 2   C.) minimum; 29

D.) maximum; 3   E.) no maximum or minimum exists.

3.) Find all solutions of $x^2 - 2x + 10 = 0$ and write the answers in the form $a + bi$.

A.) $\dfrac{1}{2} + 5i, \dfrac{1}{2} - 5i$ B.) $1 + 3i, 1 - 3i$ C.) $1 + 6i, 1 - 6i$

D.) $\dfrac{1}{2} + \sqrt{11}i, \dfrac{1}{2} - \sqrt{11}i$ E.) $2 + 3i, 2 - 3i$

4.) Solve the inequality $|2x - 3| \geq 5$ and express the solution in interval form.

A.) $x = 4$ B.) $[4, \infty)$ C.) $(-\infty, -4] \cup [4, \infty)$

D.) $(-\infty, -1] \cup [4, \infty)$ E.) $[-4, 4]$

5.) Find the domain of $g(x) = \dfrac{\sqrt{x + 13}}{x - 1}$ and express the answer in interval form.

A.) $[13, \infty)$ B.) $(-13, \infty)$ C.) $[-13, 1) \cup (1, \infty)$

D.) $(-\infty, -3) \cup (1, \infty)$ E.) $(-\infty, -1) \cup (1, \infty)$

6.) Write $\ln(5) - 4\ln(x) - 3\ln(x^2 + 7)$ as a single logarithm.

A.) $\dfrac{\log(5(x^2 + 7)^3)}{\log(x^4)}$ B.) $\ln\dfrac{5(x^2 + 7)^3}{x^4}$ C.) $\dfrac{\ln(5(x^2 + 7)^3)}{\ln(x^4)}$

D.) $\ln\dfrac{5}{x^4(x^2 + 7)^3}$ E.) $\ln\dfrac{5}{12x(x^2 + 7)}$

7.) Find the quotient (q) and remainder (r) if $x^5 + x^4 + x^2 + 1$ is divided by $x^2 + x + 1$.

A.) $q = x^3; r = x - 1$ B.) $q = x^3 - x + 2; r = -x - 1$ C.) $q = x^3; r =$

D.) $q = x^3 + 2x^2 + 3x + 6; r = 9x + 7$ E.) $q = x^2 + 2x + 3; r = 4$

8.) If $f(x) = 4x^2 - 7$, find $\dfrac{f(a + h) - f(a)}{h}$, where $h \neq 0$.

A.) 1   B.) $\dfrac{4h^2 - 14}{h}$ C.) $\dfrac{f(4h^2 - 7) - f(4a^2 - 7)}{h}$ D.) $4h$ E.) $8a + 4h$

9.) Express the equation $\log_a(b) = c$ in exponential form.

A.) $a^b = c$ B.) $c^b = a$ C.) $c^a = b$ D.) $a^c = b$ E.) $b^a = c$

10.) A 2 m tall man is walking away from a lamp post with a light source that is 8 m above the ground. How far from the lamp post is the man when his shadow is 4 m long?

A.) 12 m B.) 10 m C.) 14 m D.) 16 m E.) 15 m

11.) Find the time in years (rounded to the nearest quarter year) for an investment of \$5,000 to grow to \$60,000 at an interest rate of 13.5% compounded quarterly.

A.) 18.75 yrs   B.) 18.5 yrs   C.) 75 yrs   D.) 9.75 yrs   E.) 12 yrs

12.) Find all value(s) of k that ensure that $4x^2 + kx + 25 = 0$ has *exactly* one solution.

A.) $k = 20, -20$ B.) $k = 10$ C.) $k = 5$

D.) $k = 10, -10$ E.) There are an infinite number of solutions.

13.) If $x = 7 - 3i$ is a zero of a polynomial with integer coefficients, then which of the following is also a zero of the polynomial?

A.) $x = 7$ B.) $x = 3i$ C.) $x = 7 + 3i$

D.) $x = -7 + 3i$ E.) $x = 7 - 3i$

14.) Which of the following is the graph of $y = -\log_3(x - 2) + 1$?

A.) 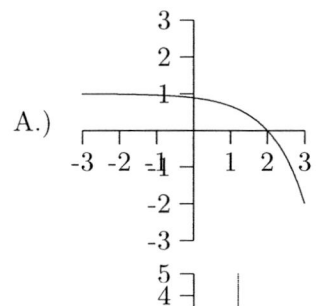   B.) 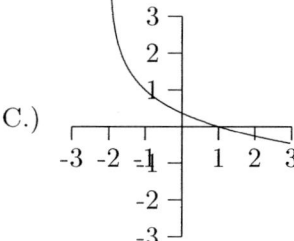   C.)

D.) 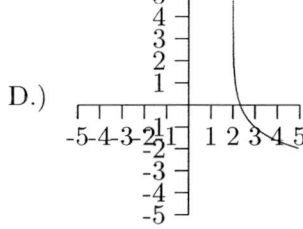   E.)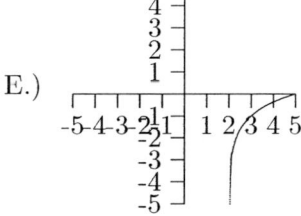

Show all work on the remaining problems in the space provided.

15.) Find all real solution(s) of the system of equations below. Your answer(s) should be in ordered pairs.

$$\begin{cases} x + \sqrt{y} = 0 \\ y^2 - 4x^2 = 12 \end{cases}$$

16.) Solve the inequality. Then express the solution in interval form and graph the solution set on the real number line.

$$1 + \frac{2}{x + 1} \le \frac{2}{x}$$

17.) a) Express the statement "s is inversely proportional to the square root t" as a formula.

b) Use the fact that if s is 25, then t is 100 to find the constant of proportionality.

c) If $s = 50$, find t.

18.) The graph of $y = f(x)$ is given. Sketch the graphs asked for in parts a) - c) on the axes provided. Include at least two ordered pairs on your final graphs.

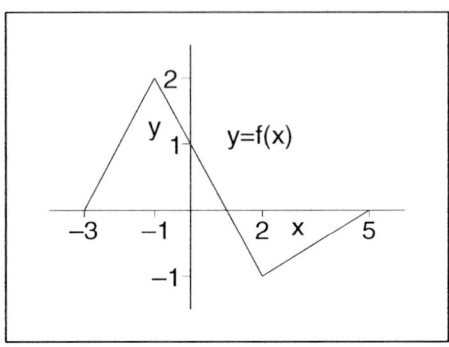

a) $y = f(x + 4)$

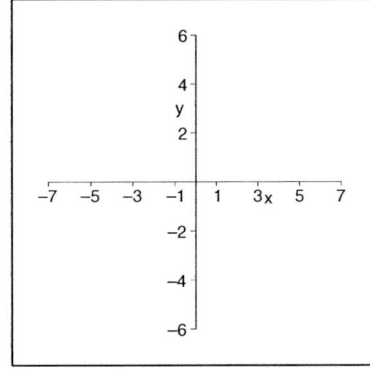

b) $y = -f(x) - 3$

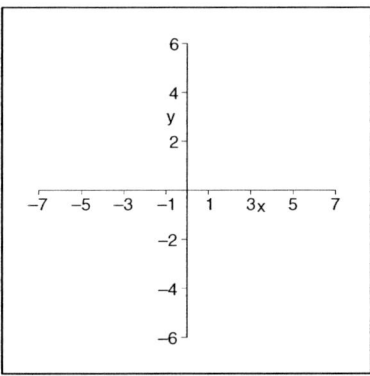

c) $y = 2f(-x)$

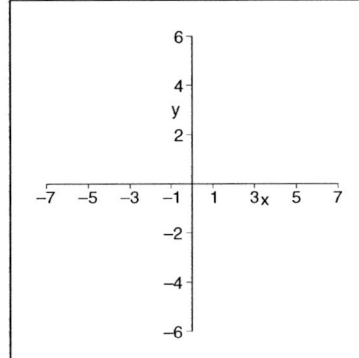

19.) Find the solution(s) (if any exist) of the equation $log_2(x + 1) = log_2(x) + 3$.

20.) If $f(x) = x^3 + 2x - 1$ and $g(x) = x^2 - 5$, find and simplify the following:

a) $(fg)(1) =$

b) $(g \circ f)(x) =$

21.) For the following questions, use $f(x) = 5x^3 - 15x^2 - 20x + 60$.
a) Is $x = 3$ an upper bound for the zeroes of f according to the upper and lower bounds theorem? If not, find the smallest integer that is an upper bound according to the upper and lower bounds theorem.

b) Is $x = -1$ a lower bound for the zeroes of f according to the upper and lower bounds theorem? If not, find the largest integer that is a lower bound according to the upper and lower bounds theorem.

c) Completely factor $f(x)$.

29.2   Spring 2003 Final

Multiple Choice (version A). Only the answer recorded on the scantron will be graded.

1.) If $f(x) = x^2 - 2$, find $f(a + 1)$.

A.) $x^2 - 1$ B.) $a^2 + 2a - 1$ C.) $a^2 - 1$

D.) $a^2 + 2a + 1$ E.) None of these.

2.) Find the exact solution to the equation $2(3^x) = 5$.

A.) $x = \log_5(6)$ B.) $x = \log_2\left(\dfrac{5}{3}\right)$ C.) $x = \log_3\left(\log_2(5)\right)$

D.) $x = \log_8(5)$ E.) $x = \dfrac{\ln\left(\frac{5}{2}\right)}{\ln(3)}$

3.) Solve the equation $|3x + 2| = 7$ for x.

A.) $x = \dfrac{5}{3}$ B.) $x = \pm\dfrac{5}{3}$ C.) $x = \dfrac{5}{3}, -3$

D.) $x = 7$ E.) None of these.

4.) Let $f(x) = \sqrt{5 + x}$ and $g(x) = 2x - 7$. Find the domain of $\left(\dfrac{f}{g}\right)(x)$.

A.) $\left[-5, \dfrac{2}{7}\right) \cup \left(\dfrac{2}{7}, \infty\right)$ B.) $[-5, \infty)$ C.) $(0, \infty)$

D.) $\left\{x \big| x \neq \dfrac{7}{2}\right\}$ E.) $\left[-5, \dfrac{7}{2}\right) \cup \left(\dfrac{7}{2}, \infty\right)$

5.) The graph of the function $y = f(x - 1) - 3$ can be obtained by shifting the graph of $y = f(x)$:

A.) One unit to the left and 3 units downward.
B.) One unit to the right and 3 units downward.
C.) 3 units to the left and 1 unit downward.
D.) 3 units to the right and 1 units upward.
E.) None of these.

6.) Find $3x - y$, if (x, y) is the solution to the system of linear equations

$$\begin{cases} 3x - 2y & = 3 \\ x + 4y & = 8. \end{cases}$$

A.) 3   B.) 4.5   C.) 8   D.) 2   E.) 1

7.) The average rate of change of the function $f(x) = x^2 - 2x + 1$ between $x = -1$ and $x = 5$ is which of the following?

A.) 1   B.) 2   C.) 6   D.) 16   E.) None of these.

8.) Let $f(x) = x^2 + 9$ and $g(x) = \sqrt{x}$. Find $(g \circ f)(4)$.

A.) $x = 13$ B.) $x = 26$ C.) $x = 5$ D.) $x = \sqrt{23}$ E.) $x = \sqrt{11}$.

9.) How many liters of a mixture containing 80% alcohol should be added to 5 liters of a 20% solution to yield a 30% solution?

A.) 1   B.) 2   C.) 3   D.) 4   E.) None of these.

10.) The remainder when the polynomial $P(x) = 3x^6 + 5x^5 - 4x + 2$ is divided by $x + 1$ is which of the following?

A.) 6   B.) 4   C.) 2   D.) 0   E.) 8.

11.) Find all real solutions of the equation:

$$\frac{1}{x - 1} + \frac{x}{x + 2} = 2$$

A.) $-1 \pm \sqrt{3}$ B.) $-2, 1$ C.) $-1 \pm \sqrt{7}$ D.) $-2, 0, 1$ E.) None of these.

12.) The solution to the inequality $\dfrac{12}{x - 5} \leq -2$ is

A.) $[-1, 5)$ B.) $(5, \infty)$ C.) $(-\infty, -1) \cup (5, \infty)$ D.) $(-\infty, 5)$ E.) $(-\infty, 3)$.

13.) Solve the equation $\log(x - 1) + \log(x + 2) = 1$ for x

A.) $x = -3$ B.) $x = -4, 3$ C.) $x = -\frac{7}{3}$ D.) $x = 5.5$ E.) None of these.

14.) In order to divide $\dfrac{3 + 5i}{1 - 2i}$ correctly, the first step should be to multiply the numerator and denominator by which of the following?

A.) $1 - 2i$ B.) $3 - 5i$ C.) $-1 - 2i$ D.) $1 + 2i$ E.) $-3 + 5i$

Show all work on the remaining problems in the space provided to receive credit.

15.) a) Show that the equation $x^2 + y^2 - 2x + 4y + 1 = 0$ represents a circle by putting the circle in standard form.
b) Find the center (C) of the circle.

c) Find the radius (R) of the circle.

16.) Find a polynomial P with integer coefficients so that P has degree 3 and zeroes -3 and $1 + i$.

17.) Find the equation of the line in slope-intercept form that passes through $(-\frac{1}{2}, 3)$ and is perpendicular to the line $y = -\frac{2}{5}x + 3$.

18.) Find all solutions to the system of equations

$$\begin{cases} y + x^2 = 4x \\ y + 4x = 16. \end{cases}$$

19.) Let $f(x) = -x^2 + 4x + 5$.

a. Put $f(x)$ into the standard form of $f(x) = a(x - h)^2 + k$. Be sure to show the "complete the square" steps.

b. Find the x- and y- intercepts.

c. Graph the function on the axes below.

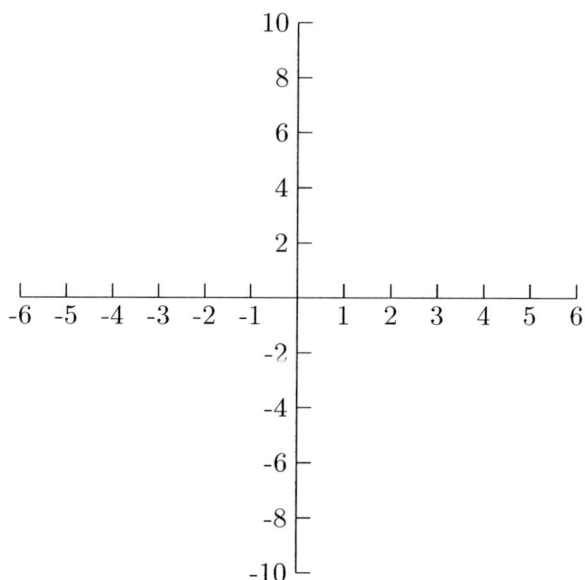

20.) Suppose s is jointly proportional to t and u, and is inversely proportional to x^2.

a.) Express the above relationship as a formula, using k as the constant of proportionality.

b.) Suppose that when $x = 1$, $t = 2$, and $u = 3$, then $s = 2$. Use this information to find the constant of proportionality k.

c.) Using your answers from a.) and b.), find s when $x = 2$, $t = 1$, and $u = 3$.

21.) For the equation $x^3 - x^2 + 3x - 60$, do each of the following:

a.) Find all possible rational solutions provided by the Rational Zeros Theorem.

b.) Find one rational solution *using synthetic division.*

c.) Use your answer in part b.) to find the remaining solutions.

29.3   Spring 2004 Final

1. Which of the following lines is parallel to the line through the points $(-3, 4)$ and $(1, 2)$?

   A.) $x + 2y = 6$ B.) $-3x + 4y = 2$ C.) $x - 2y = 7$
   D.) $-x + 2y = 6$ E.) $2x + y = 5$

2. $(12, -4)$ is a point on the circle $(x - 5)^2 + (y + 4)^2 = 49$. How far is $(12, -4)$ from the center of the circle?

   A.) 7   B.) 8   C.) $\sqrt{145}$ D.) 49   E.) None of these.

3. Find the range of $y = 6^{(x-3)} + 2$.

   A.) $(2, \infty)$ B.) $(-\infty, 3)$ C.) $(-\infty, 2) \cup (2, \infty)$
   D.) $(-\infty, 2) \cup (2, 3) \cup (3, \infty)$ E.) $[3, \infty)$

4. Which of the following is the graph of a function that is odd?

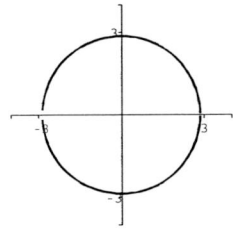 B.) 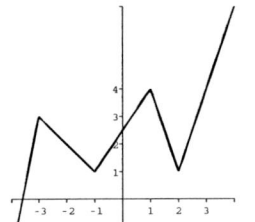)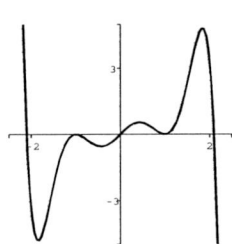

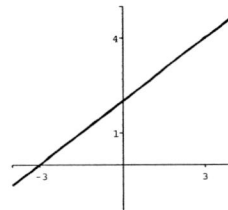 E.)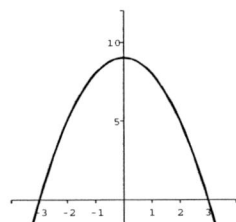

5. If $\log_b(5) = 3$, then what is b?

   A.) $b = 3^5$ B.) $b = \sqrt[5]{3}$ C.) $b = \sqrt[3]{5}$
   D.) $b = 5^3$ E.) None of these

6. Find the solutions of $6x - 2x^2 = 5$.

   A.) $\dfrac{3}{2} \pm i$ 　 B.) $\dfrac{3}{2} \pm \dfrac{1}{2}i$ 　 C.) $-6 \pm \dfrac{1}{2}i$ 　 D.) $\dfrac{-3}{2} \pm \dfrac{\sqrt{19}}{2}$ 　 E.) $\dfrac{5}{2}, -1$

7. Which of the following represents a mathematical model of the statement F varies directly as g and inversely as r squared (here k represents a nonzero constant)?

   A.) $F = kgr^2$ 　　　　 B.) $F = k\dfrac{g}{\sqrt{r}}$ 　　　　 C.) $F = kg\sqrt{r}$

   D.) $F = k\dfrac{r^2}{g}$ 　　　　 E.) $F = k\dfrac{g}{r^2}$

8. If $\lfloor x \rfloor$ represents the Greatest Integer Function of x and $g(x) = \lfloor x \rfloor$, what is the value of $g(-1.25) + g(4.75) - g(2)$?

   A) 1 　　 B.) 3 　　 C.) 2.5 　　 D.) 0 　　 E.) 2

9. Rewrite $\log_2(x^4 + 5) - 3\log_2(z) + 6\log_2(y+1)$ as one logarithm.

   A.) $\log_2\left(\dfrac{2(x^4 + 5)(y+1)}{z}\right)$ 　　　　　　　　 B.) $\left(\dfrac{(x^4 + 5)(y+1)^6}{z^3}\right)$

   C.) $\dfrac{1}{2}\log_2\left(\dfrac{(x^4 + 5)z}{y+1}\right)$ 　　　　　　　　 D.) $\log_2\left(\dfrac{(x^4 + 5)(z)}{2(y+1)}\right)$

   E.) $\log_2\left(\dfrac{(x^4 + 5)(y+1)^6}{z^3}\right)$

10. Find the domain of $(f/g)(x)$ if $f(x) = \sqrt{x+4}$ and $g(x) = x^2 - 1$.

    A.) $[-4, -1) \cup (1, \infty)$ 　　　　 B.) $[-4, \infty)$ 　　　　 C.) $[-4, 1) \cup (1, \infty)$
    D.) $(4, \infty)$ 　　　　　　　　　　 E.) None of these.

11. For what values of k will $kx^2 + 10x + k = 0$ have *exactly* one solution?

    A.) $k = 10$ 　　　　　　 B.) $k = 5$ 　　　　　　 C.) $k = 25$
    D.) $k = \pm 5$ 　　　　　 E.) None of these.

12. Find the constant c so that the denominator of $\dfrac{x^3 - 4x^2 + 7x + c}{x - 3}$ will divide evenly into the numerator?

A.) -12 B.) 3   C.) 84   D.) 7   E.) -30

13. Suppose $f(x) = x^2 - 3x$. Find and simplify $\dfrac{f(a + h) - f(a)}{h}$ for $h \neq 0$.

A.) $h - 3$ B.) 1   C.) $2a + h - 3$ D.) $2a + h + 3$ E.) $2a + h^2 - 3h$

14. Given the graphs of $f(x)$ and $g(x)$ below, find $(f \circ g)(2)$.

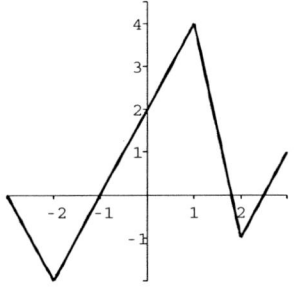

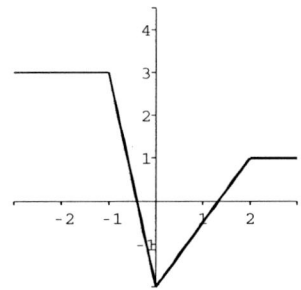

Figure 29.1: Graph of $y = f(x)$ Figure 29.2: Graph of $y = g(x)$

A.) -1 B.) 3   C.) 2   D.) 1   E.) 4

15. Which of the following is the solution of $|2x - 3| > 5$?

A.) $(4, \infty)$ B.) $(-1, 4)$ C.) $(1, 4)$
D.) $(-\infty, -1) \cup (4, \infty)$ E.) None of these.

Show all work on the remaining problems in the space provided to receive credit.

16. Ahmed plans to invest his $8,000 into two different bank accounts for one year. One account pays 6% simple interest and the other account pays 3% simple interest. Ahmed wants to earn a total of $375 in interest. How much money should he invest into the account that pays 6% interest?

17. The rate at which rumors spread around Big U. is given by
    $N = \dfrac{25000}{1 + 100e^{-.05t}}$, where N is the number of students that have heard the rumor and t is the time in days.

    (a) How many students at Big U. have heard the rumor after 10 days? Round to the nearest student.

    (b) How many days will it be before 18,000 students at Big U. have heard the rumor? Round to the nearest day.

18. Sketch the graph of the solution of the system of inequalities below. Label the vertices.

$$\begin{cases} 3x + 2y & \le 6 \\ x & \le 1 \\ x & \ge 0 \\ y & \ge 0 \end{cases}$$

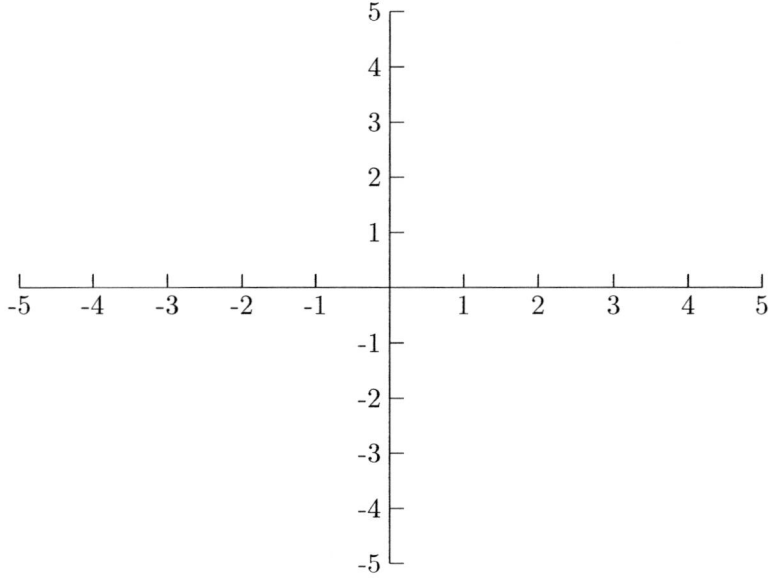

19. The graph of $y = h(x)$ was obtained by transforming the graph of $y = |x|$. Find $h(x)$ and record your answer in the space provided below. Your answer should be an equation.

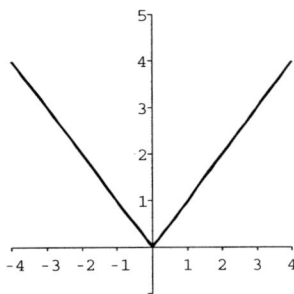

Figure 29.3: Graph of $y = |x|$

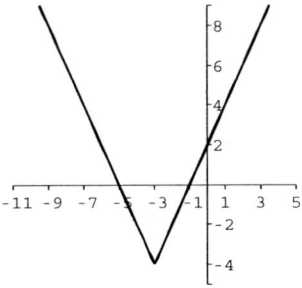

Figure 29.4: Graph of $y = h(x)$

$h(x) = $ _____

20. Find all real solutions of $\dfrac{2}{x} = \dfrac{3}{x-2} - 1$.

21. Use $f(x) = 2x^3 - 2x^2 - 74x - 70$ to answer the questions that follow.

    (a) According to the Rational Zeros Test, what are the possible rational zeros? List them all in the space provided.

    (b) Use the Remainder Theorem to show that $x = -1$ is a root of $f(x)$.

    (c) Completely factor $f(x)$.

29.4 Spring 2005 Final

Multiple Choice (version A). Only the answer recorded on the scantron will be graded.

1. Suppose that the line with equation $y = mx + b$ passes through the points $(7, 4)$ and $(-13, 4)$. Then $m + b =$

   A.) $1/5$ B.) -5 C.) 0 D.) -4 E.) 4

2. Find the equation of a line perpendicular to $2x - 3y = 6$ that passes through the point $(4, -5)$.

   A.) $y = -\dfrac{3}{2}x + 1$ B.) $y = \dfrac{3}{2}x - 5$ C.) $y = \dfrac{3}{2}x - 11$

   D.) $y = -\dfrac{2}{3}x - \dfrac{7}{3}$ E.) $y = \dfrac{2}{3}x - \dfrac{23}{3}$

3. What is the domain of $f(x) = \dfrac{x}{x^2 - 1}$?

   A.) $(-\infty, -1) \cup (-1, \infty)$ B.) $(-\infty - 1) \cup (-1, 0) \cup (0, 1) \cup (1, \infty)$
   C.) $(-\infty - 1) \cup (-1, 1) \cup (1, \infty)$ D.) All real numbers.
   E.) $x = 1$ or $x = -1$

4. $\frac{1}{3}\log 8 - 2\log 4$ is equivalent to

   A.) $\log \dfrac{1}{8}$ B.) $-\dfrac{2}{3}\log 2$ C.) $\log 16$ D.) $-\dfrac{5}{3}\log 2$ E.) $\dfrac{1}{3}\log_{16} 8$

5. Suppose that $x - y = 2$ and $2x - 3y = 1$. Then $x + y =$

   A.) 2 B.) 8 C.) $1/3$ D.) 9 E.) 3

6. What is the solution for $|2x - 1| > 7$?

   A.) $(-\infty, -3) \cup (4, \infty)$ B.) $(-3, \infty) \cup (4, \infty)$ C.) $(-3, 4)$
   D.) $(4, -3)$ E.) $(4, \infty)$

7. Where will the point $(3, 7)$ on the graph of $y = f(x)$ get shifted to on the graph of $y = 2f(x - 1) + 4$?

   A.) $(2, 10)$ B.) $(4, 6)$ C.) $(10, 8)$ D.) $(4, 18)$ E.) $(2, 18)$

8. Floyd cycles at 25 mph and Levi cycles at 20 mph. When they start cycling directly towards each other, they are 90 miles away from each other. How far has Floyd traveled when Floyd and Levi meet?

   A.) 45 miles   B.) 90 miles   C.) 50 miles   D.) 40 miles   E.) 22.5 miles

9. If $x = -2 + 3i$ is a zero of a polynomial with integer coefficients, what else is?

   A.) $x = -3i$ B.) $x = 2 - 3i$ C.) $x = \sqrt{13}$
   D.) $x = -2 - 3i$ E.) $x = 2 + 3i$

10. Solve the logarithmic equation $\log_2(\log(x)) = -1$ for x.

    A.) $\sqrt{10}$ B.) 5   C.) 1/100   D.) x^{-20} E.) $\log(-1/5)$

11. Evaluate $\log_5(110)$ using the change of base formula. Round to two decimal places.

    A.) 4.70   B.) 2.92   C.) 2.04   D.) 0.34   E.) 22.00

12. Given the graphs of $y = f(x)$ and $y = g(x)$ below, find $(f \circ g)(2)$.

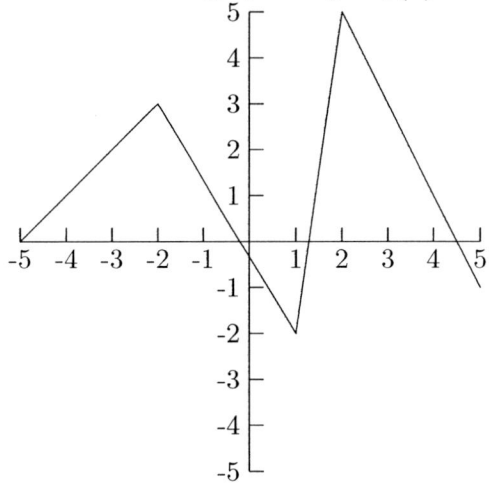

$$y = f(x)$$

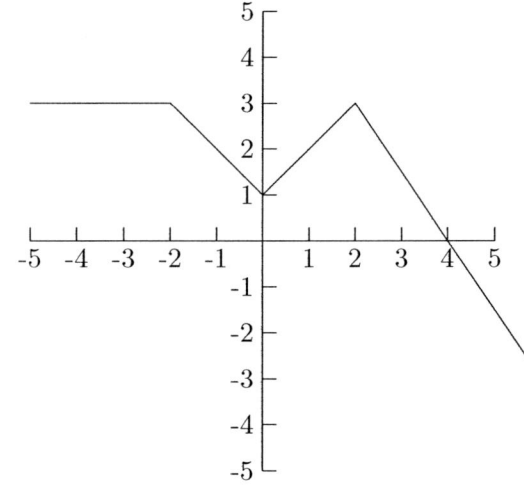

$$y = g(x)$$

    A.) -1 B.) 3   C.) 2   D.) -2 E.) 5

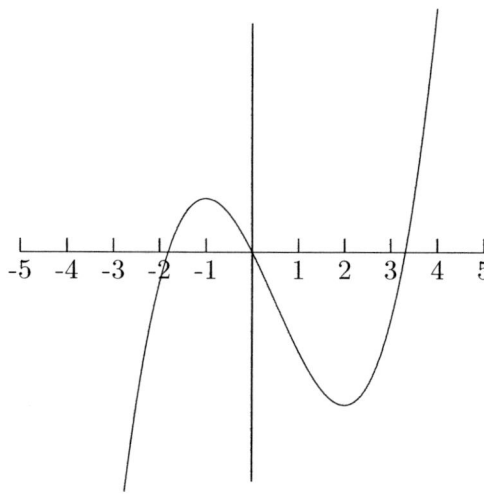

13. For the function whose graph is above, where is the function increasing?

A.) $(-1, 2)$ B.) $(-\infty, -1) \cup (2, \infty)$ C.) $(-2, 1)$

D.) $(-\infty, -2) \cup (1, \infty)$ E.) The function is never increasing.

14. The graph below is of the form $f(x) = 2^{x-a} + b$. Find a and b.

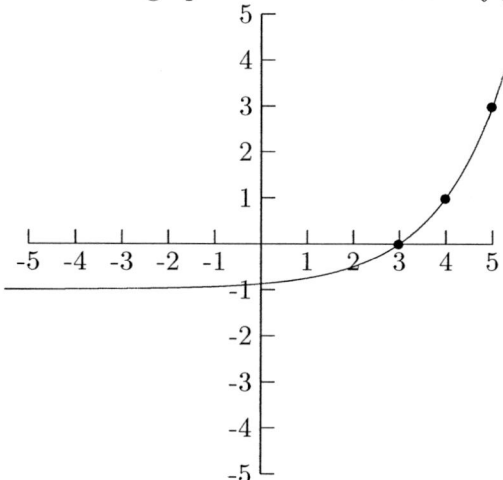

A.) $a = -3, b = 1$ B.) $a = 3, b = -1$ C.) $a = 2, b = -1$

D.) $a = 1, b = -4$ E.) $a = 1, b = 2$

15. Which of the following points is a solution to the system of inequalities below?

$$\begin{cases} x + y \leq 1 \\ 2x - 3y \geq 2 \end{cases}$$

A.) $(0,0)$ B.) $(1,1)$ C.) $(-10,0)$ D.) $(2,5)$ E.) $(1,-1)$

Show all work on the remaining problems in the space provided to receive credit.

16. Solve $\sqrt{2x+1} + 1 = x$ for x.

17. Solve the equation $1000e^{(0.05t)} = 2,300,000$ for t. Your solution should involve the use of logarithms. No credit will be given for work that is not shown. Round your answer to two decimal places.

18. Find a quadratic function that has vertex $(1,3)$ and that contains the point $(-2,8)$. Put your answer in the form $f(x) = a(x-h)^2 + k$.

19. $f(x) = x^3 + 4x^2 + 3x - 2.$

    (a) List the possible rational zeros of f.

    (b) Find the rational zero. (*Hint:* You may want to use synthetic division.)

    (c) Write $f(x)$ as a product of a linear term and a quadratic term.

    (d) Find the other two zeros of f. Simplify your solutions. Decimal approximations are not acceptable.

20. If $f(x) = 3x^2 - 5.$

    (a) Evaluate $f(2)$.

    (b) Simplify $f(a + 1)$.

    (c) Simplify $\dfrac{f(x + h) - f(x)}{h}$.

21. Solve $7 \geq 8 - 2x > -3$. Write your answer in interval notation.

29.5   Fall 2005 Final

1. Find the x and y intercepts of the graph of the equation $2x - 4y + 16 = 0$.

   A.) $(2, 0)$ and $(0, -4)$ B.) $(8, 0)$ and $(0, -4)$ C.) $(-2, 0)$ and $(4, 0)$
   D.) $(-8, 0)$ and $(0, 4)$ E.) $(-2, 0)$ and $(0, 4)$

2. Which of the following is the equation of the circle with radius 3 and center $(4, -4)$ in standard form?

   A.) $(x - 4)^2 + (y + 4)^2 = 9$ B.) $(x + 4)^2 + (y - 4)^2 = 9$
   C.) $(x - 4)^2 + (y + 4)^2 = 3$ D.) $(x + 4)^2 + (y - 4)^2 = 3$
   E.) $(x - 3)^2 + (y - 3)^2 = 16$

3. Which of the following is the equation of the line containing the point $(1, -2)$ and perpendicular to the graph of $2x + y = 5$?

   A.) $y + 2 = \dfrac{1}{2}(x - 1)$ B.) $y + 2 = -2(x - 1)$ C.) $y - 2 = -\dfrac{1}{2}(x + 1)$
   D.) $y - 2 = \dfrac{1}{2}(x + 1)$ E.) $y + 2 = 2(x - 1)$

4. You are driving on a Canadian freeway to a town that is 200 kilometers from your home. After 30 minutes you pass an exit that you know is 50 kilometers from your home. Assuming you continue at a constant speed, how long will it take for the entire trip?

   A.) 1 hours and 30 minutes   B.) 1 hours and 45 minutes   C.) 2 hours
   D.) 2 hours and 15 minutes   E.) 2 hours and 30 minutes

5. Find all of the values of k that ensure $x^2 + kx + 9 = 0$ has exactly one real solution.

   A.) $x = 1$ B.) $x = \pm 6$ C.) $x = 3$ D.) $x = 0$ E.) $x = -9$

6. Solve the inequality $6 \leq 2x + 10 < 12$ for x.

   A.) $[-2, 1)$ B.) $(-1, 2]$ C.) $(1, -2]$ D.) $[6, 12)$ E.) $[2, 6)$

7. Solve the inequality $|3 - 4x| \leq 7$ for x.

   A.) $[-1, \infty)$ B.) $(-\infty, 5/2) \cup [1, \infty)$ C.) $[-1, 5/2]$

   D.) $[-5/2, 1]$ E.) $[1, 2]$

8. Find the domain of $g(x) = \dfrac{\sqrt{x + 5}}{x^2 - 4}$.

   A.) $[-5, \infty)$ B.) $[-5, -2) \cup (-2, 2) \cup (2, \infty)$

   C.) $(2, \infty)$ D.) $(\infty, -2) \cup (2, \infty)$

   E.) $[-5, -2) \cup (2, \infty)$

Use the graphs below to answer questions 9 - 10.

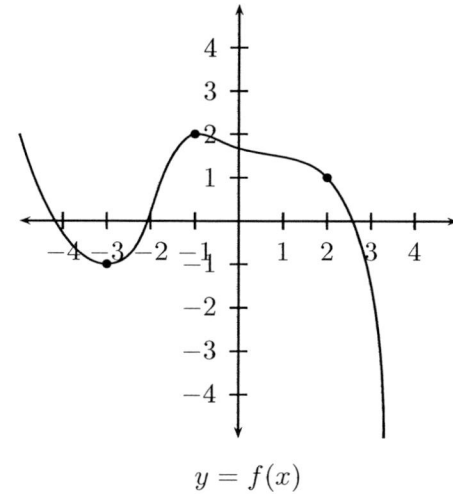
$$y = f(x)$$

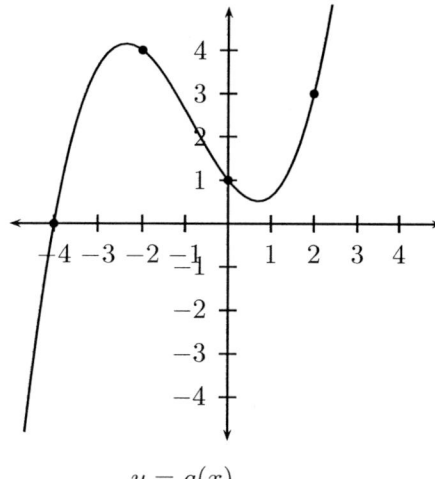
$$y = g(x)$$

9. On the graph of $y = f(x)$ above, where is the function decreasing?

   A.) $(4, \infty)$ B.) $(-3, -1)$ C.) $(-\infty, -1) \cup (3, \infty)$

   D.) $(-\infty, -3) \cup (-1, 2) \cup (2, \infty)$ E.) $(-\infty, -3) \cup (-1, \infty)$

10. Using the graphs of $y = f(x)$ and $y = g(x)$ above, find $(g - f)(2)$.

   A.) -1 B.) 0 C.) -2 D.) 2 E.) 1

11. The graph of $y = h(x)$ (on the right below) is obtained by transforming the graph of $y = |x|$ (on the left below). Find a formula for $h(x)$.

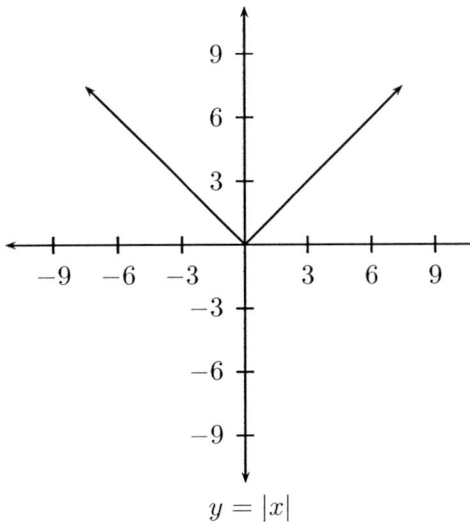

$y = |x|$

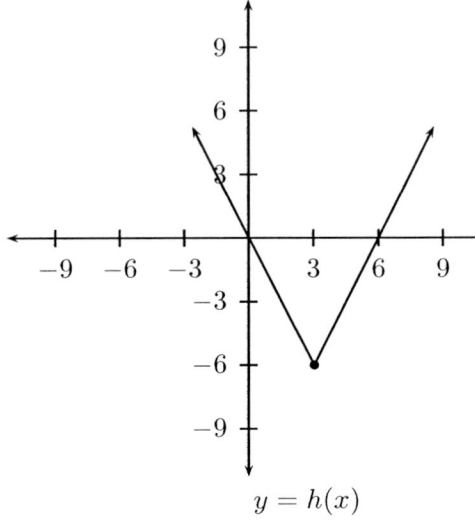

$y = h(x)$

A.) $h(x) = |x + 3| - 6$ B.) $h(x) = 2|x - 3| - 6$ C.) $h(x) = |2(x - 6)| + 3$

D.) $h(x) = |x - 3| + 6$ E.) $h(x) = |1/2x - 3| - 6$

12. Rewrite $x^\pi = e$ in logarithmic form.

A.) $\log_\pi(x) = e$ B.) $\ln(x) = \pi$ C.) $\log_x(\pi) = e$

D.) $\log_\pi(e) = x$ E.) $\log_x(e) = \pi$

13. Use the Laws of logarithms to write $2\left[\log_3 x - \dfrac{1}{2}\log_3(x + 2) - 4\log_3(x - 5)\right]$ as a single logarithm.

A.) $\log_3\left(\dfrac{x^2(x - 5)^8}{x + 2}\right)$ B.) $\log_3\left(\dfrac{x^2}{\sqrt{x + 2}\,(x - 5)^4}\right)$

C.) $\log_3\left(\dfrac{x}{(x + 2)(x - 5)}\right)$ D.) $2\log_3\left(\dfrac{x}{\sqrt{x + 2}\,(x - 5)^4}\right)$

E.) $\log_3\left(\dfrac{x^2}{(x + 2)(x - 5)^8}\right)$

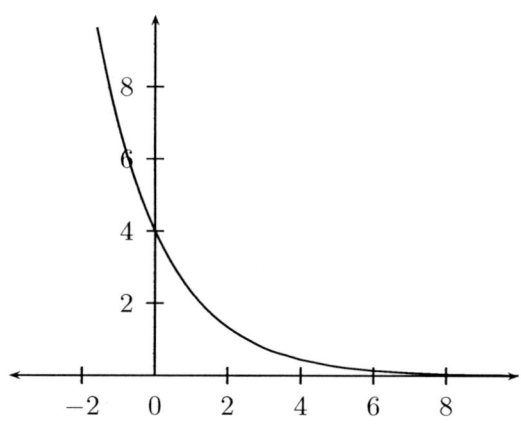

14. If $a > 1$, $0 < b < 1$ and $c > 0$, then the graph above could be which of the following?

A.) $y = c \cdot a^x$ B.) $y - c \cdot b^x$ C.) $y = c \cdot \log_a(x)$

D.) $y = c \cdot \log_b(x)$ E.) $y = c \cdot x^a$

15. Suppose that P is directly proportional to x and the square of y and inversely proportional to z. Express the relationship above as a formula using k as the constant of proportionality.

A.) $P = \dfrac{x \cdot y^2}{z}$ B.) $P = k \ \dfrac{x \cdot \sqrt{y}}{z}$ C.) $P = k \ \dfrac{x \cdot y^2}{z}$

D.) $P = \dfrac{x \cdot \sqrt{y}}{z}$ E.) $P = k \ \dfrac{x}{y^2 \cdot z}$

16. Find the exact solution of $7(4^x) = 15$

A.) $x = \log_{15}(4)$ B.) $x = \dfrac{\log(\frac{15}{7})}{4}$ C.) $x = \log_4(\log_7(15))$

D.) $x = \log_{2401}(15)$ E.) $x = \dfrac{\log(\frac{15}{7})}{\log 4}$

Show all work on the remaining problems in the space provided to receive credit.

17. Let $h(x) = |3 + 5x^4|$.

(a) If $f(x) = |x|$, find $g(x)$ so that $h(x) = (f \circ g)(x)$.

(b) If $g(x) = x^4$, find $f(x)$ so that $h(x) = (f \circ g)(x)$.

18. Use the function $f(x)$ given below to answer the questions that follow.

$$f(x) = \begin{cases} 1 & \text{if } x \leq -2 \\ x - 1 & \text{if } 0 < x < 1 \\ 1 - \sqrt{x} & \text{if } x > 1 \end{cases}$$

(a) $f(-2) =$ _____

(b) $f(4) =$ _____

(c) Sketch the graph of $y = f(x)$ on the axes provided below.

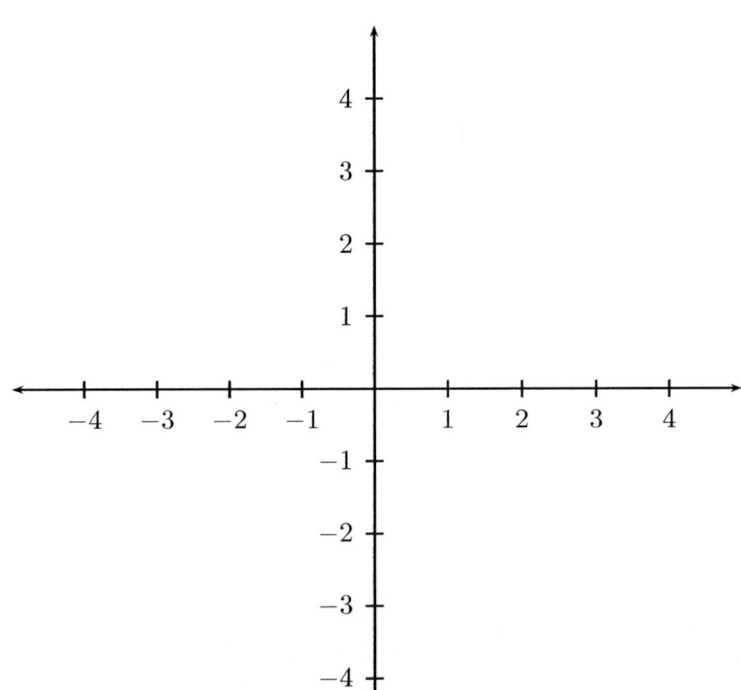

19. Use $g(x) = x^4 - 3x^3 + 5x^2 - x - 10$ to answer the questions that follow.

    (a) According to the Rational Zeros Theorem, what are the possible rational zeros of $g(x)$? List them all in the space provided.

    (b) $x = 2$ is a zero of $g(x)$ and $g(x)$ has one other rational zero. Find the other rational zero of $g(x)$.

    (c) Use synthetic division to help write $g(x)$ as a product of two linear terms and an irreducible quadratic term.

    (d) Use the fact that $x = 1 + 2i$ is a zero to completely factor $g(x)$ over the complex numbers.

20. First Bank is advertising that its savings account offers continuously compounded interest at a rate that would allow \$300 to grow to \$500 in only 4 years. Find the continuously compounded interest rate r that First Bank is offering. Write your answer as a percentage rounded to two decimal places.

21. Solve the following system using either substitution or elimination. (Circle which method was used).

    Substitution   **Elimination**

    $$\begin{cases} 2x + y = 1 \\ 3x + 2y = 4 \end{cases}$$

22. Determine which system of inequalities that made the graph below.

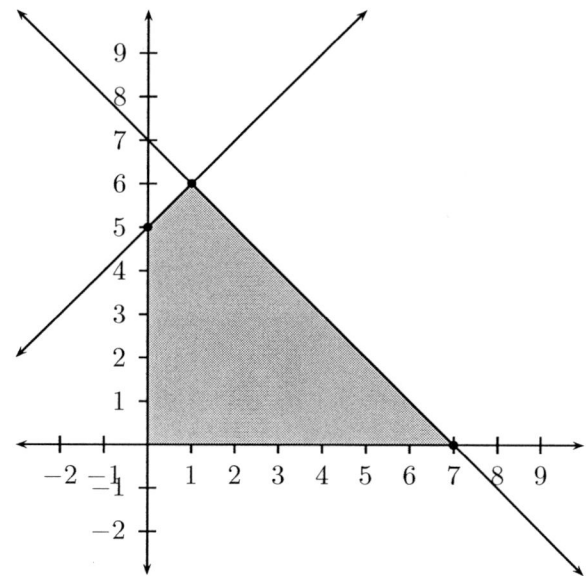

29.6   Spring 2006 Final

1. Which of the following is an equation of the line through the points $(2, 1)$ and $(-3, 4)$?

   A.) $-x + 2y - 11 = 0$ B.) $3x + 5y + 1 = 0$ C.) $2x + y - 1 = 0$
   D.) $x - y - 1 = 0$ E.) $3x + 5y - 11 = 0$

2. Find the equation of the line perpendicular to $x = -7$ that passes through the point $(-6, 1)$.

   A.) $y = 1$ B.) $x = 1$ C.) $y = -6$ D.) $y = -7$ E.) $x = -6$

3. Which of the answers below is equivalent to $\sqrt{-10}\sqrt{-5}$?

   A.) $\sqrt{50}$ B.) $5\sqrt{-2}$ C.) $5\sqrt{2}$ D.) $(i\sqrt{10})(i\sqrt{5})$ E.) 50

4. Which of the answers below gives the correct solution to $x^2 + 5x = 10$?

   A.) $x = \dfrac{-5 \pm i\sqrt{15}}{2}$ B.) $x = 10, 5$ C.) $x = -5 \pm \sqrt{35}$

   D.) $x = \dfrac{-5 \pm \sqrt{65}}{2}$ E.) $x = -5 \pm \sqrt{10}$

5. Which of the following is the solution of $\dfrac{1}{x - 2} \le \dfrac{1}{x^2 - 4}$?

   A.) $(-\infty, -2) \cup [-1, 2)$ B.) $(-\infty, -1]$ C.) $(-2, -1]$
   D.) $[-1, \infty)$ E.) $(-2, -1] \cup (2, \infty)$

6. Evaluate $f(a + h) - f(a)$ if $f(x) = x^2 - 4$.

   A.) h^2 B.) $2ah + h^2$ C.) $2ah + h^2 - 8$
   D.) $ah + h^2$ E.) h

7. Which of the following is equivalent to $a^c = b$?

   A.) $\log_a b = c$ B.) $\log_c a = b$ C.) $\log_c b = a$
   D.) $\log_a c = b$ E.) $\log_b a = c$

8. Which of the following could not be a rational zero of $P(x) = 24x^{17} - 6x^{10} + 5x^2 - 9$ according to the Rational Zeros Theorem?

   A.) $1/6$ B.) $-1/4$ C.) 9 D.) $-3/8$ E.) $2/3$

Use the graph below to answer questions 9.

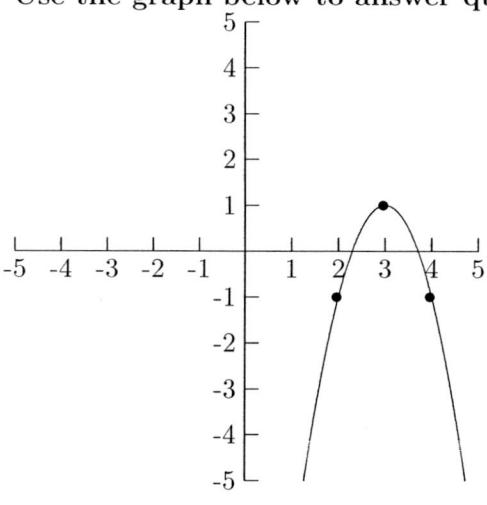

9. The graph above is of the form $y = a(x - b)^2 + c$. Find a, b and c.

   A.) $a = 2, b = 3, c = -1$ B.) $a = -2, b = 3, c = 1$

   C.) $a = -2, b = -3, c = -1$ D.) $a = -1, b = 3, c = 1$

   E.) $a = -1, b = -3, c = 1$

10. Find the domain of $g(x) = \log_5(2x + 7)$.

   A.) $(0, \infty)$ B.) $(7/2, \infty)$ C.) $(-7/2, \infty)$

   D.) $(-\infty, 0) \cup (0, \infty)$ E.) $(-7/2, 0) \cup (0, \infty)$

Use the graphs below to answer question 11.

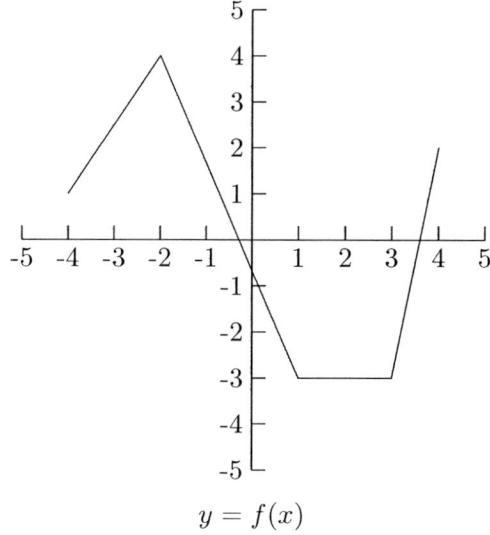

$y = f(x)$

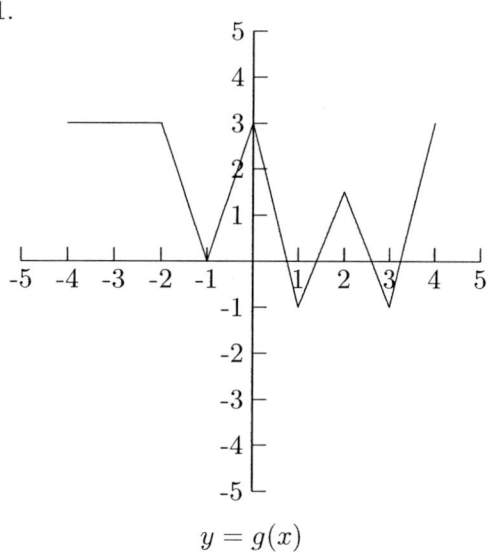

$y = g(x)$

11. Find $(fg)(1)$.

   A.) 4   B.) 3   C.) -1 D.) 2   E.) -3

12. Solve $\ln(x) - \ln(x - 4) = 5$ for x.

    A.) 5   B.) $\dfrac{4e^5}{e^5 - 1}$ C.) 5, 1   D.) $4 + \dfrac{\ln(3) - 1}{5}$ E.) e^5

13. Which of the following correctly describes the end behavior of $P(x) = -x^5 + 4x^4 - x^3 + 5$?

    A.) As $x \to -\infty, y \to -\infty$ B.) As $x \to -\infty, y \to -\infty$
        As $x \to \infty, y \to -\infty$ As $x \to \infty, y \to \infty$

    C.) As $x \to -\infty, y \to \infty$ D.) As $x \to -\infty, y \to \infty$
        As $x \to \infty, y \to -\infty$ As $x \to \infty, y \to \infty$
    E.) None of the above.

14. If $x + 2$ is a factor of $f(x) = x^3 - 4x^2 + kx - 6$, then what would the value of k have to be?

    A.) 0   B.) -15 C.) -2 D.) 30   E.) 2

15. The graph below represents the solution of which of the following system of inequalities?

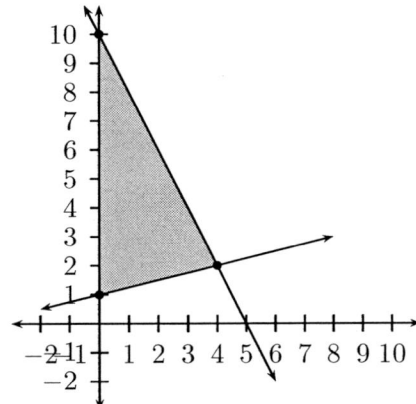

    A.) $\begin{cases} y \geq x + 1 \\ y \leq x - 10 \\ x \geq 0 \\ y \geq 0 \end{cases}$ B.) $\begin{cases} y \leq \frac{1}{4}x + 1 \\ y \geq -2x + 10 \\ x \geq 0 \\ y \geq 0 \end{cases}$ C.) $\begin{cases} y \geq \frac{1}{4}x + 1 \\ y \leq -2x + 10 \\ x \geq 0 \\ y \geq 0 \end{cases}$

    D.) $\begin{cases} y \geq \frac{1}{4}x + 1 \\ y \geq -2x + 10 \\ x \geq 0 \\ y \geq 0 \end{cases}$ E.) $\begin{cases} y > \frac{1}{4}x + 1 \\ y < -2x + 10 \\ x \geq 0 \\ y \geq 0 \end{cases}$

Show all work on the remaining problems in the space provided to receive credit.

16. Sarah sells a fruit juice that she makes by mixing together fruit juice concentrate and water. She wants her mixture to contain 15 liters of 20% strength juice. How much of the fruit juice concentrate will she need if the concentrate contains 80% juice?

17. Alicia is saving for retirement and she wants to know how long it will take her to have $1,000,000 saved. In the equation below, n represents the number of years that it will take for Alicia to save $1,000,000 if she saves $12,000 per year and earns 9% interest compounded annually.
Find n (rounded up to the nearest whole number).

$$1000000 = 12000 \left(\frac{1.09^n - 1}{.09} \right)$$

18. Use the graph of $y = g(x)$ below to sketch the graph of $y = 2g(-x) + 3$ on the provided set of axes.

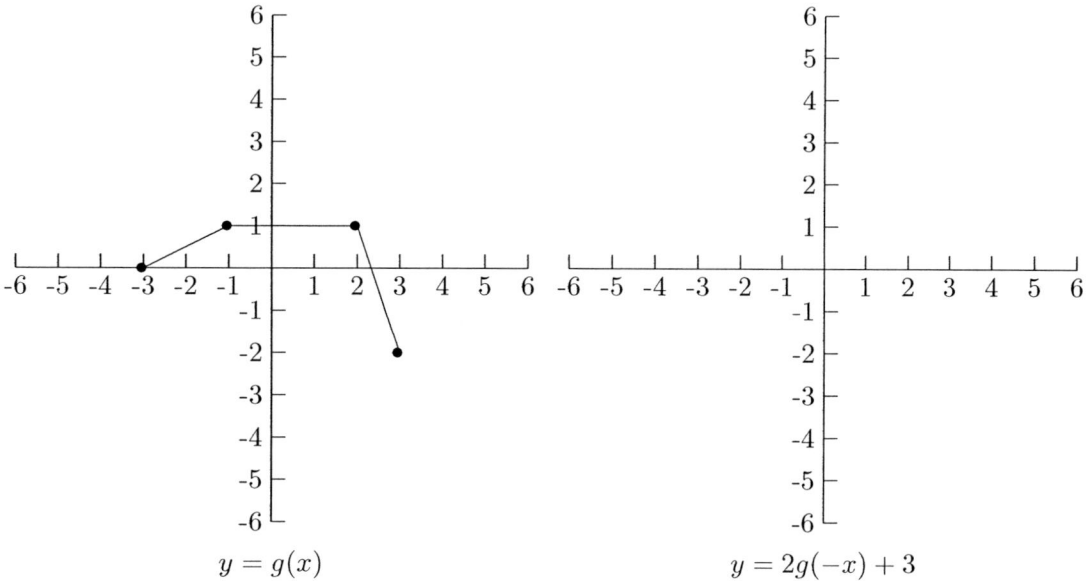

$y = g(x)$ $y = 2g(-x) + 3$

19. Use the function $f(x)$ given below to answer the questions that follow.

$$f(x) = \begin{cases} x + 3 & \text{if } x \leq -1 \\ 3 & \text{if } -1 < x < 3 \\ -(x - 4)^3 & \text{if } 3 \leq x \leq 5 \end{cases}$$

(a) $f(-2) =$ _____ .

(b) $f(0) =$ _____ .

(c) $f(f(5)) =$ _____ .

(d) Sketch the graph of $y = f(x)$ on the axes provided below.

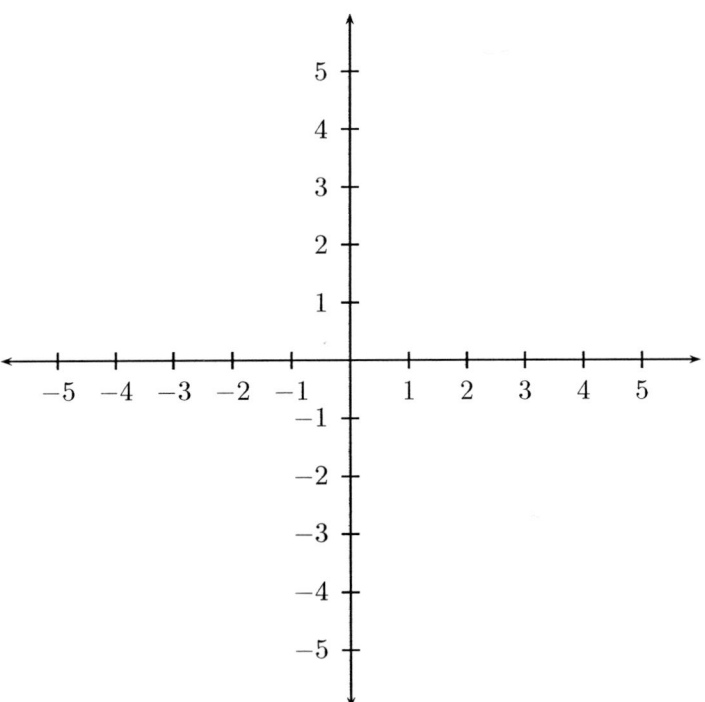

20. Solve $-|3x - 7| + 10 < 8$. Put your answer in interval notation.

21. Solve the following system using either substitution or elimination. (Circle which method was used).

   Substitution   **Elimination**

$$\begin{cases} 2x + 4y & = -4 \\ 5x - 7y & = 41 \end{cases}$$

22. For this problem, $f(x) = 3x^3 - 10x^2 + x + 14$.

   (a) Use synthetic division to show that $x + 1$ is a factor of $f(x)$.

   (b) Use the remainder theorem to show that $x - 2$ is a factor of $f(x)$.

   (c) Completely factor $f(x)$.

29.7   Cover page for Spring 2004 exam

This page is provided to show you what a typical Final Exam cover page looks like.

NAME (Print) _____

Student number _____

Section number _____

Instructor's name _____

MATH 1100 FINAL EXAM

May 4th, 2004

This exam consists of two sections. The first section contains 15 multiple choice problems. **Only the answers recorded on your scantron will be graded,** so please carefully bubble in the answers on your scantron. Make sure that your name, your instructor's name, and your section number are recorded on your scantron. The scantron must be filled in using a number 2 pencil. Each multiple choice problem is worth 4 points.

The second section of the exam consists of 6 free response problems. You must show your work to receive any credit and only work shown on the exam itself will be graded. Please transfer any and all work that you want graded from your scratch paper to the exam itself. The number of points each free response problem is worth is listed below in parentheses next to each problem's number.

No notes are allowed for use on the exam. Only approved calculators are allowed for use on the exam. Good luck!

Do not write below. For grading purposes only.

16. _____ (7)

17. _____ (8)

18. _____ (6)

19. _____ (6)

20. _____ (7)

21. _____ (7)

Multiple Choice:

number correct _____ x 4 = _____ (60)

Total _____ (101)

Chapter 30

Solutions of Old Finals

30.1 Fall 2002 Final Solutions

1. The slope of the line passing through the points $(2, 5)$ and $(-2, 1)$ is
$$m = \frac{1 - 5}{-2 - 2} = \frac{-4}{-4} = 1.$$
The equation of the line with slope $m = 1$ and passing through the point $(1, 7)$ is $y - 7 = 1(x - 1) \Rightarrow y = x + 6$.
Therefore the correct answer is **D**.

2. Since $a = 3 > 0$, the graph of $f(x)$ has a minimum. To find the vertex of $f(x)$, let's write it in the standard form $f(x) = a(x - h)^2 + k$.
$f(x) = 3x^2 - 18x + 29$ factor 3 from the first two terms, we have
$f(x) = 3(x^2 - 6x) + 29 = 3(x^2 - 6x + 3^2) + 29 - 3 \cdot 3^2 = 3(x - 3)^2 + 2$
Therefore, the vertex is $(3, 2)$ and the minimum is 2.
The answer is **B**.

3. Solving this quadratic equation by completing the square gives
$x^2 - 2x + 10 = 0 \Rightarrow x^2 - 2x = -10 \Rightarrow$
$x^2 - 2x + 1 = -9 \Rightarrow (x - 1)^2 = -9 \Rightarrow x - 1 = \pm\sqrt{-9} \Rightarrow x = 1 \pm 3i$
The answer is **B**.

4. $|2x - 3| \geq 5 \Rightarrow 2x - 3 \leq -5$ or $2x - 3 \geq 5 \Rightarrow$
$2x \leq -2$ or $2x \geq 8 \Rightarrow x \leq -1$ or $x \geq 4$. The solution is $(-\infty, -1] \cup [4, \infty)$.
The answer is **D**.

5. The domain of $g(x) = \dfrac{\sqrt{x + 13}}{x - 1}$ is the set of all x for which
$x + 13 \geq 0$ and $x - 1 \neq 0$, that is the set of all x where $x \geq -13$ and $x \neq 1$.
So the domain of g is $[-13, 1) \cup (1, \infty)$.
The answer is **C**.

6. Using the properties of logarithms, we can write
$\ln 5 - 4 \ln x - 3 \ln(x^2 + 7) = \ln 5 - \ln x^4 \ln(x^2 + 7)^3 =$
$\ln \dfrac{5}{x^4} - \ln (x^2 + 7)^3 = \ln \dfrac{5/x^4}{(x^2 + 7)^3} = \ln \dfrac{5}{x^4(x^2 + 7)^3}$.
The correct answer is **D**.

7. The long division is shown below.

$$
\begin{array}{r}
 x^3 -x + 2 \\
\hline
x^2 + x + 1\,)\quad x^5 + x^4 + 0x^3+ x^2 + 0x + 1 \\
\underline{-(x^5 + x^4 + x^3)} \\
-x^3 + x^2 + 0x + 1 \\
\underline{-(-x^3 - x^2 - x\,)} \\
2x^2 + x + 1 \\
\underline{-(2x^2 + 2x + 2\,)} \\
-x - 1
\end{array}
$$

Thus, the quotient is $x^3 - x + 2$ and the remainder is $-x - 1$.

The correct answer is **B**.

8. Since $f(x) = 4x^2 - 7$,
$f(a) = 4a^2 - 7$, and $f(a + h) = 4(a + h)^2 - 7 =$
$4(a^2 + 2ah + h^2) - 7 = 4a^2 + 8ah + 4h^2 - 7$. Therefore,
$$
\frac{f(a + h) - f(a)}{h} = \frac{4a^2 + 8ah + 4h^2 - 7 - (4a^2 - 7)}{h} =
$$
$$
\frac{8ah + 4h^2}{h} = \frac{h(8a + 4h)}{h} = 8a + 4h.
$$
The answer is **E**.

9. $\log_a b = c \Rightarrow b = a^c$.
The answer is **D**.

10. $\dfrac{4 + x}{8} = \dfrac{4}{2} \Rightarrow 4 + x = 16 \Rightarrow x = 12$.
The answer is **A**.

11. $A = P\left(1 + \dfrac{r}{n}\right)^{nt}$. Here we have $A = 60,000$, $P = 5,000$, $r = 0.135$, and
$n = 4$
To find t, we solve the equation
$$
60,000 = 5,000\left(1 + \frac{0.135}{4}\right)^{4t} \Rightarrow \left(1 + \frac{0.135}{4}\right)^{4t} = 12 \Rightarrow
$$
$$
\ln\left(1 + \frac{0.135}{4}\right)^{4t} = 12 \Rightarrow 4t\left(1 + \frac{0.135}{4}\right) = 12 \Rightarrow t = \frac{\ln 12}{4\ln\left(1 + \frac{0.135}{4}\right)}
$$

≈ 18.72. years
The answer is **A**.

12. The equation $4x^2 + kx + 25 = 0$ has exactly one solution when
$b^2 - 4ac = 0 \Rightarrow k^2 - 4 \cdot 4 \cdot 25 = 0 \Rightarrow$
$k^2 - 400 = 0 \Rightarrow (k - 20)(k + 20) = 0$
$k = 20$ or $k = -20$.
The answer is **A**.

13. If $7 - 3i$ is a zero of a polynomial with integer coefficients, then its conjugate $7 + 3i$ is also a zero.
The answer is **C**.

14. The graph of $y = -\log_3(x - 2) + 1$ should be the graph of $y = \log_3(x)$ reflected across the x-axis, shifted right 2 units and up 1 unit.
The answer is **B**.

15. To solution shown uses the substitution method.
From the first equation we have $x + \sqrt{y} = 0 \Rightarrow x = -\sqrt{y}$.
Replace x with its value in terms of y in the second equation.
$y^2 - 4(-\sqrt{y})^2 = 12 \Rightarrow y^2 - 4y - 12 = 0 \Rightarrow (y - 6)(y + 2) = 0$
$\Rightarrow y = 6$ or $y = -2$.
Since $x = -\sqrt{y}$, y cannot be negative. Therefore, $y = 6$ and $x = -\sqrt{6}$.
The solution is the ordered pairs $(-\sqrt{6}, 6)$.

16. $1 + \dfrac{2}{x+1} \le \dfrac{2}{x} \Rightarrow 1 + \dfrac{2}{x+1} - \dfrac{2}{x} \le 0 \Rightarrow \dfrac{x(x+1) + 2x - 2(x+1)}{x(x+1)} \le 0$
$\dfrac{x^2 + x + 2x - 2x-}{x(x+1)} \le 0 \Rightarrow \dfrac{x^2 + x - 2}{x(x+1)} \le 0 \Rightarrow \dfrac{(x-1)(x+2)}{x(x+1)} \le 0$
The critical numbers for this inequality are -2, -1, 0, and 1.

Intervals	$(-\infty, -2]$	$[-2, -1)$	$(-1, 0)$	$(0, 1]$	$[1, \infty)$
Testing Point	$x = -3$	$x = -\dfrac{3}{2}$	$x = -\dfrac{1}{2}$	$x = \dfrac{1}{2}$	$x = 2$
$\dfrac{(x-1)(x+2)}{x(x+1)}$	$\dfrac{(-4)(-1)}{(-3)(-2)} = \dfrac{2}{3}$	$-\dfrac{5}{3}$	9	$-\dfrac{5}{3}$	$\dfrac{2}{3}$
Signs of $\dfrac{(x-1)(x+2)}{x(x+1)}$	$+$	$-$	$+$	$-$	$+$

The solution is S $= [-2, -1) \cup (0, 1]$

17.

a. $S = \dfrac{k}{\sqrt{t}}$, where k is a constant.

b. S = 25, when $t = 100 \Rightarrow 25 = \dfrac{k}{\sqrt{100}} \Rightarrow k = 250$.

c. Since S $= \dfrac{250}{\sqrt{t}}$ and S = 50, then $50 = \dfrac{250}{\sqrt{t}} \Rightarrow \sqrt{t} = 5 \Rightarrow t = 25$.

18.

(a)   (b)

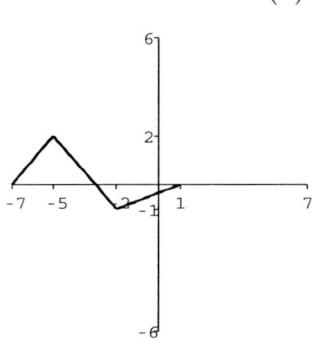
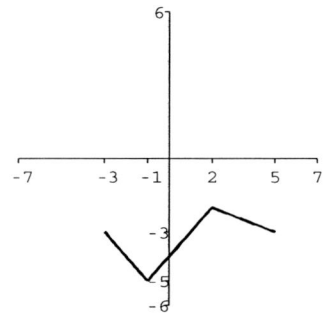

Figure 30.1: $y = f(x+4)$.

Figure 30.2: $y = -f(x) - 3$

(c)

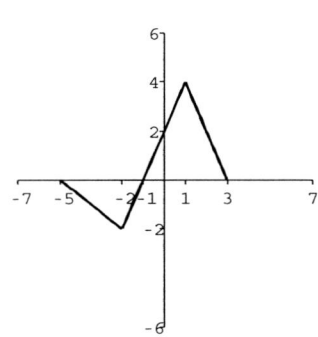

Figure 30.3: $y = 2f(-x)$

19. $\log_2(x+1) = \log_2 x + 3 \Rightarrow \log_2(x+1) - \log_2 x = 3 \Rightarrow$
$\log_2(\frac{x+1}{x}) = 3 \Rightarrow \frac{x+1}{x} = 2^3 \Rightarrow 8x = x+1 \Rightarrow 7x = 1 \Rightarrow x = \frac{1}{7}$

20.

a. $(f \cdot g)(1) = f(1) \cdot g(1) = (1^3 + 2 \cdot 1 - 1)(1^2 - 5) = 2(-4) = -8$

b. $(g \circ f)(x) = g(f(x)) = g(x^3 + 2x - 1) = (x^3 + 2x - 1)^2 - 5.$

21.

3	5	-15	-20	60
		15	0	-60
	5	0	-20	0

a. $x = 3$ is not a an upper bound for the zeroes of $f(x)$, since the numbers in the third row of the division of $f(x)$ by $(x - 3)$ are not all nonnegative.
$x = 4$ is an upper bound for the zeroes of $f(x)$, since the numbers in the third row of the division of $f(x)$ by $(x - 4)$ are all nonnegative.

4	5	-15	- 20	60
		20	20	0
	5	5	0	60

b. $x = -1$ is not a lower bound for the zeros of $f(x)$, since the numbers in the third row of the division of $f(x)$ by $(x + 1)$ are not alternating.

-1	5	-15	- 20	60
		-5	20	0
	5	-20	0	60

$x = -2$ is the largest integer that is a lower bound for the zeros of $f(x)$, since the numbers in the third row of the division of $f(x)$ by $(x + 2)$ are alternating

-2	5	-15	- 20	60
		-10	50	-60
	5	-25	30	0

c. From part (a) 3 is a zero of $f(x)$. Therefore,
$f(x) = (x - 3)(5x^2 - 20) = 5(x - 3)(x^2 - 4) = 5(x - 3)(x - 2)(x + 2).$

30.2 Spring 2003 Final Solutions

1. If $f(x) = x^2 - 2$, then $f(a+1)^2 - 2 = a^2 + 2a + 1 - 2 = a^2 + 2a - 1$. The correct answer is **B**.

2. $2 \cdot 3^x = 5 \Rightarrow 3^x = \dfrac{5}{2} \Rightarrow \ln 3^x = \ln \dfrac{5}{2} \Rightarrow$

   $x \cdot \ln 3 = \ln \dfrac{5}{2} \Rightarrow x = \dfrac{\ln \frac{5}{2}}{\ln 3}$. The answer is **E**.

3. $|3x + 2| = 7 \Rightarrow$

   If $3x + 2 = -7$, then $3x = -9 \Rightarrow x = -3$.

   If $3x + 2 = 7$, then $3x = 5 \Rightarrow x = \dfrac{5}{3}$.

   The correct answer is **C**.

4. The domain of $\left(\dfrac{f}{g}\right)(x) = \dfrac{\sqrt{x+5}}{2x-7}$ is the set of all x where $x + 5 \geq 0$ and $2x - 7 \neq 0$.

   That is the set of all x where $x \geq -5$ and $x \neq \dfrac{7}{2}$, so the domain of f is

   $\left[-5, \dfrac{7}{2}\right) \cup \left(\dfrac{7}{2}, \infty\right)$ which is **E**.

5. To graph $y = f(x-1) - 3$, we shift the graph of $y = f(x)$ one unit to the right and three units downward. The answer is **B**.

6. To find $3x - y$, we need to solve the system. To do that we can multiply the first equation by 2 and eliminate the y.

   $2 \cdot (3x - 2y = 3) \Rightarrow 6x - 4y = 6$. Adding this to the second equation gives $7x = 14 \Rightarrow x = 2$

   Since $x = 2$, we can use the 2nd equation to find y.

   $2 + 4y = 8 \Rightarrow 4y = 6 \Rightarrow y = \dfrac{6}{4} = \dfrac{3}{2}$.

   Therefore, $3x - y = 3 \cdot 2 - \dfrac{3}{2} = 6 - 1.5 = 4.5$ and the answer is **B**.

7. The average rate of change of $f(x) = x^2 - 2x + 1$, between $x = -1$ and $x = 5$ is

   $\dfrac{f(5) - f(-1)}{5 - (-1)} = \dfrac{5^2 - 2 \cdot 5 + 1 - ((-1)^2 - 2 \cdot 1 + 1)}{5 + 1} =$

   $\dfrac{25 - 10 + 1 - (1 + 2 + 1)}{6} = \dfrac{16 - 4}{6} = \dfrac{12}{6} = 2$.

   The answer is **B**.

8. $(g \circ f)(4) = g(f(4)) = g(4^2 + 9) = g(16 + 9) = g(25) = \sqrt{25} = 5$

The answer is **C**.

9. Let $x =$ the amount of 80 percent mixture, then

$.8x + .2(5) = .3(x + 5) \Rightarrow .8x + 1 = .3x + 1.5 \Rightarrow .5x = .5 \Rightarrow x = 1$ Liter.

The answer is **A**.

10. The remainder
$R = P(-1) = 3(-1)^4 + 5(-1)^3 - 4(-1) + 2 = 3 - 5 + 4 + 2 = 4$. The answer is **B**.

11. To solve the equation $\dfrac{1}{x-1} + \dfrac{x}{x+2} = 2$, we multiply both sides of the equation by the LCD of $(x-1)$ and $(x+2)$, which $(x-1)(x+2)$, obtaining
$x + 2 + x(x-1) = 2(x-1)(x+2) \Rightarrow x + 2 + x^2 - x = 2(x^2 + x - 2) \Rightarrow$
$x^2 + 2 = 2x^2 + 2x - 4 \Rightarrow x^2 + 2x - 6 = 0$.

That's a quadratic equation in x. We can solve this equation by completing the square.

$x^2 + 2x - 6 = 0 \Rightarrow x^2 + 2x = 6 \Rightarrow x^2 + 2x + 1 = 6 + 1 \Rightarrow$

$(x + 1)^2 = 7 \Rightarrow x + 1 = \pm\sqrt{7} \Rightarrow x = -1 \pm \sqrt{7}$.

The answer is **C**.

12. $\dfrac{12}{x-5} + 2 \le 0 \Rightarrow \dfrac{12 + 2(x-5)}{x-5} \le 0 \Rightarrow \dfrac{12 + 2x - 10}{x-5} \le 0 \Rightarrow \dfrac{2x+2}{x-5} \le 0$.

The critical numbers of the inequality -1 and 5 divide the real line into three intervals $(-\infty, -1], [-1, 5)$, and $(5, \infty)$. Note that we did not put a bracket around 5, since 5 makes the denominator zero. To determine where $\dfrac{2x+2}{x-5} \le 0$ we pick a test point in each of the intervals.

The critical numbers for this inequality are -1 and 5.

Intervals	$(-\infty, -1]$	$[-1, 5)$	$(5, \infty)$
Testing Point	$x = -2$	$x = 0$	$x = 10$
$\dfrac{2x+2}{x-5}$	$\dfrac{2(-2)+2}{(-2)-5} = \dfrac{2}{7}$	$\dfrac{2(0)+2}{(0)-5} = -\dfrac{2}{5}$	$\dfrac{2(10)+2}{(10)-5} = \dfrac{22}{5}$
Signs of $\dfrac{2x+2}{x-5}$	$+$	$-$	$+$

Thus $\dfrac{2x+2}{x-5} \le 0$, when x is in $[-1, 5)$. Therefore the solution is $S = [-1, 5)$.

The answer is **A**.

13.

$$\begin{aligned}
\log(x-1) + \log(x+2) = 1 &\Rightarrow \log(x-1)(x+2) = 1 \Rightarrow \\
(x-1)(x+2) &= 10^1 \Rightarrow \\
x^2 + x - 2 &= 10 \Rightarrow \\
x^2 + x - 12 &= 0 \Rightarrow \\
(x-3)(x+4) &= 0 \Rightarrow x = 3 \text{ or } x = -4.
\end{aligned}$$

Since $x = -4$ makes $\log(x-1)$ undefined, the only solution is $x = 3$.
The answer is **E**.

14. In order to divide $\dfrac{3+5i}{1-2i}$, we have to multiply the numerator and denominator by the complex conjugate of the denominator $1 - 2i$. The complex conjugate is $1 + 2i$.
The answer is **D**.

15. (a) $x^2 + y^2 - 2x + 4y + 1 = 0 \Rightarrow x^2 - 2x + y^2 + 3y = -1 \Rightarrow$
$x^2 - 2x + 1 + y^2 + 4y + 4 = -1 + 1 + 4 \Rightarrow (x-1)^2 + (y+2)^2 = 4$.

(b) The center is at $(1, -2)$.

(c) The radius is 2.

16. The factor theorem says that c is a zero of $p(x) \iff (x - c)$ is a factor of $p(x)$.
Since $1 + i$ is a zero of $p(x)$, its conjugate $1 - i$ is also a zero of $p(x)$. Now $(x+3)$, $(x - (1+i))$, and $(x - (1-i))$ are factors of $p(x)$. Therefore,

$$\begin{aligned}
p(x) = (x+3)(x-(1+i))(x-(1-i)) &= (x+3)(x-1-i)(x-1+i) \\
&= (x+3)(x^2 - 2x + 2) \\
&= x^3 - 2x^2 + 2x + 3x^2 - 6x + 6 \\
&= x^3 + x^2 - 4x + 6.
\end{aligned}$$

17. The slope of the line that is perpendicular to the line $y = \dfrac{-2}{5}x + 4$ is given by $m = -\dfrac{1}{\frac{2}{5}} = \dfrac{5}{2}$. The equation of the line with slope $m = \dfrac{5}{2}$ and passing through the point $\left(3, -\dfrac{1}{2}\right)$ is

$$y - \left(\dfrac{-1}{2}\right) = \dfrac{5}{2}(x-3) \Rightarrow y + \dfrac{1}{2} = \dfrac{5}{2}x - \dfrac{15}{2} \Rightarrow y = \dfrac{5}{2} - 8.$$

18. To use the substitution method, we will solve the first equation for y in terms of x. This yields $y + x^2 = 4x \Rightarrow y = 4x - x^2$. Replacing y in the second equation, we have $(4x - x^2) + 4x = 16$. Solving for x gives $(4x - x^2) + 4x = 16 \Rightarrow x^2 - 8x + 16 = 0 \Rightarrow (x - 4)^2 = 0 \Rightarrow x = 4$. Back-substituting to find y gives $y = 4(4) - (4)^2 = 0$. Thus, the only solution is $(4, 0)$.

19. (a) $f(x) = -x^2 + 4x + 5 = -(x^2 - 4x + \ ) + 5 = -(x^2 - 4x + 4) + 5 + 4 = -(x - 2)^2 + 9$.

    (b) To find the x-intercepts, set $f(x) = 0$ and solve for x.
    $-x^2 + 4x + 5 = 0 \Rightarrow x^2 - 4x - 5 = 0 \Rightarrow (x - 5)(x + 1) = 0 \Rightarrow x = 5$ or $x = -1$.
    Thus, the x-intercepts are $(5, 0)$ and $(-1, 0)$.
    Alternatively, $-(x - 2)^2 + 9 = 0 \Rightarrow 9 = (x - 2)^2 \Rightarrow \pm 3 = x - 2 \Rightarrow x = 2 \pm 3 \Rightarrow x = 5$ or $x = -1$.

    (c) The graph is below.

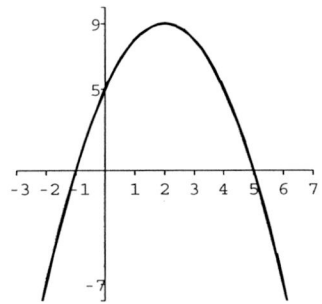

Figure 30.4: $y = f(x) = -(x - 2)^2 + 9$.

20. (a) $s = k\dfrac{t \cdot u}{x^2}$.

    (b) $s = 2$ when $x = 1, t = 2$, and $u = 3 \Rightarrow 2 = k\dfrac{2 \cdot 3}{1^2} \Rightarrow k = \dfrac{2}{6} = \dfrac{1}{3}$.

    (c) When $x = 2, t = 1$ and $u = 3$, $s = \dfrac{1}{3} \cdot \dfrac{t \cdot u}{x^2} = \dfrac{1}{3} \cdot \dfrac{1 \cdot 3}{2^2} = \dfrac{1}{4}$.

21. (a) The possible rational zeros of $x^3 - x^2 - 5x + 6 = 0$ are the factors of the constant coefficients 6 over the factors of the leading coefficient 1. That is $\dfrac{\text{the factors of } 6}{\text{the factors of } 1} = \dfrac{\pm 1, \pm 2, \pm 3, \pm 6}{\pm 1} = \pm 1, \pm 2, \pm 3, \pm 6.$

    (b) The synthetic division below shows that $x = 2$ is a zero of $P(x) = x^3 - x^2 - 5x + 6.$

$$
\begin{array}{r|rrrr}
2 & 1 & -1 & -5 & 6 \\
  &   & 2 & 2 & -6 \\
\hline
  & 1 & -1 & -3 & 0
\end{array}
$$

    (c) From part (b), we can conclude $x^3 - x^2 - 5x + 6 = (x - 2)(x^2 + x - 3).$ To find the remainder of the zeros can be found by solving $x^2 + x - 3 = 0.$ Using the quadratic formula yields $\dfrac{-b \pm \sqrt{b^2 - 4ac}}{2a} = \dfrac{-1 \pm \sqrt{1^2 - 4 \cdot 1 \cdot (-3)}}{2(1)} = \dfrac{-1 \pm \sqrt{1 + 12}}{2} = \dfrac{-1 \pm \sqrt{13}}{2}.$

    Thus the zeros are 2 and $\dfrac{-1 \pm \sqrt{13}}{2}.$

30.3   Spring 2004 Final Solutions

1. The slope between $(-3, 4)$ and $(1, 2)$ is $\dfrac{2-4}{1-^-3} = \dfrac{-2}{4} = \dfrac{-1}{2}$.

   The only line among the answers that has a slope of $\frac{-1}{2}$ is **A**.

   The correct answer is **A**.

2. By definition, every point on a circle is a distance of 1 radius from the center. The radius in this case is $\sqrt{49} = 7$.

   The correct answer is **A**.

3. The graph of $y = 6^{x-3} + 2$ is the graph of $y = 6^x$ shifted up 2 units and right 3 units. Because $y = 6^x$ has a range of $(0, \infty)$ and $y = 6^{x-3} + 2$ is shifted up 2 units, $y = 6^{x-3} + 2$ has a range $(2, \infty)$.

   The correct answer is **A**.

4. To be the graph of an odd function, the graph must pass the vertical line test and must be symmetric about the origin. The only graph fitting this description is **C**.

   The correct answer is **C**.

5. $\log_b(5) = 3 \Rightarrow b^3 = 5 \Rightarrow b = \sqrt[3]{5}$.

   The correct answer is **C**.

6. $6x - 2x^2 = 5 \Rightarrow 2x^2 - 6x + 5 = 0 \Rightarrow$
   $$x = \frac{-(^-6) \pm \sqrt{(-6)^2 - 4(2)(5)}}{2(2)} = \frac{6 \pm \sqrt{-4}}{4} = \frac{6 \pm 2i}{4} = -\frac{3}{2} \pm \frac{1}{2}i.$$

   The correct answer is **B**.

7. The problem describes $F = k\dfrac{g}{r^2}$.

   The correct answer is **E**.

8. If $g(x) = \lfloor x \rfloor$, then $g(-1.25) = -2$, $g(4.75) = 4$ and $g(2) = 2$. Thus, $g(-1.25) + g(4.75) - g(2) = -2 + 4 - 2 = 0$.

   The correct answer is **D**.

9.

   $$\log_2(x^4 + 5) - 3\log_2(z) + 6\log(y+1) = \log_2(x^4 + 5) + \log_2(y+1)^6 - \log_2(z^3) =$$
   $$\log_2\left((x^4 + 5)(y+1)^6\right) - \log_2(z^3) =$$
   $$\log_2\left(\frac{(x^4 + 5)(y+1)^6}{z^3}\right)$$

   The correct answer is **E**.

10. The $\sqrt{x+4}$-term implies that $x + 4 \geq 0 \Rightarrow x \geq -4$. The $x^2 - 1$-term in the denominator means that $x^2 - 1 \neq 0 \Rightarrow x \neq 1$ or $x \neq -1$. Thus, the domain is $[-4, -1) \cup (-1, 1) \cup (1, \infty)$.

The correct answer is **E**.

11. For $kx^2 + 10x + k = 0$ to have exactly one solution, the discriminant $b^2 - 4ac$ must be 0. Thus, $10^2 - 4(k)(k) = 0 \Rightarrow 100 - 4k^2 = 0 \Rightarrow$ $(10 - 2k)(10 + 2k) = 0 \Rightarrow 2k = 10$ or $2k = -10 \Rightarrow k = 5$ or $k = -5$.

The correct answer is **D**.

12. For $x - 3$ to divide evenly into $x^3 - 4x^2 + 7x + c$, the remainder must be zero. The synthetic division yields

$$
\begin{array}{r|rrrr}
3 & 1 & -4 & 7 & c \\
  &   & 3 & -3 & 12 \\
\hline
  & 1 & -1 & 4 & c + 12
\end{array}
$$

We see that the remainder is $c + 12$. If $x - 3$ divides evenly into $x^3 - 4x^2 + 7x + c$, then the remainder must be 0. Thus, $c + 12 = 0 \Rightarrow c = -12$.

The correct answer is **A**.

13.

$$\frac{f(a+h) - f(a)}{h} = \frac{\left((a+h)^2 - 3(a+h)\right) - (a^2 - 3a)}{h} =$$
$$\frac{a^2 + 2ah + h^2 - 3a - 3h - a^2 + 3a}{h} =$$
$$\frac{2ah + h^2 - 3h}{h} =$$
$$\frac{h(2a + h - 3)}{h} = 2a + h - 3 \text{ for } h \neq 0.$$

The correct answer is **C**.

14. From Figure 29.2, $g(2) = 1$. From Figure 29.1, $f(1) = 4$. Thus, $f(g(2)) = f(1) = 4$.

The correct answer is **E**.

15. $|2x - 3| > 5 \Rightarrow$
$2x - 3 > 5$ or $2x - 3 < -5 \Rightarrow$
$2x > 8$ or $2x < -2 \Rightarrow$
$x > 4$ or $x < -1$. This is equivalent to $(-\infty, -1) \cup (4, \infty)$.

The correct answer is **D**.

16. If x represents the amount of money going into the account earning 6%, then $8,000 - x$ remains. Thus, $8,000 - x$ must be the amount being deposited into the account earning 3%. Then $.06x$ and $.03(8000 - x)$ represent the amounts of interest earned in the two accounts, respectively. Thus, $.06(x) + .03(8000 - x) = 375 \Rightarrow .06x + 240 - .03x = 375 \Rightarrow .03x = 135 \Rightarrow x = 4500$.

17. (a) $N(10) = \dfrac{25000}{1 + 100e^{-0.05(10)}} \approx 405.49$ or 405 students.

(b) Substituting in $N = 18,000$ and solving gives

$$18000 = \frac{25000}{1 + 100e^{-0.05(t)}} \Rightarrow$$

$$18000(1 + 100e^{-0.05t}) = 25000 \Rightarrow$$

$$1 + 100e^{-0.05t} = \frac{25000}{18000} \Rightarrow$$

$$100e^{-0.05t} = \frac{25000}{18000} - 1 \Rightarrow$$

$$e^{-0.05t} = \frac{\frac{25000}{18000} - 1}{100} \Rightarrow$$

$$-0.05t = \ln\left(\frac{\frac{25000}{18000} - 1}{100}\right) \Rightarrow$$

$$t = \frac{\ln\left(\frac{\frac{25000}{18000} - 1}{100}\right)}{-0.05} \approx 110.99 \text{ or } 111 \text{ days.}$$

18. The solution of the system

$$\begin{cases} 3x + 2y \le 6, \\ x \le 1 \\ x \ge 0, \\ y \ge 0. \end{cases}$$

is below.

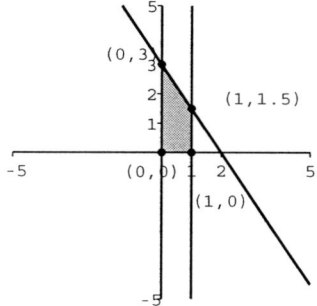

Figure 30.5: Symmetric about x-axis

19. Looking at the graph of $y = h(x)$, you will notice that the graph of $y = |x|$ has been shifted to the left 3 units and down 4 units. You should also notice that the graph is 'thinner'. This can be taken two ways: either the graph is stretched vertically by a factor of 2 or shrunk horizontally by a factor of $1/2$. Thus, either $h(x) = 2|x + 3| - 4$ or $h(x) = |2(x + 3)| - 4$.

20.

$$\frac{2}{x} = \frac{3}{x - 2} - 1 \Rightarrow 2(x - 2) = 3(x) - 1(x(x - 2)) \Rightarrow \text{ for } x \ne 0, 2$$

$$2x - 4 = 3x - x^2 + 2x \Rightarrow$$

$$x^2 - 3x - 4 = 0 \Rightarrow$$

$$(x - 4)(x + 1) = 0 \Rightarrow x = -1 \text{ or } x = 4.$$

21. (a) $\pm 1, \pm 2, \pm 5, \pm 7, \pm 10, \pm 14, \pm 35, \pm 70, \pm\dfrac{1}{2}, \pm\dfrac{5}{2}, \pm\dfrac{7}{2}, \pm\dfrac{35}{2}.$

   (b) $f(-1) = 2(-1)^3 - 2(-1)^2 - 74(-1) - 70 = -2 - 2 + 74 - 70 = 0.$
       Because $f(-1) = 0$, the Remainder Theorem, states that -1 is a root of $f(x)$.

   (c) Knowing -1 is a zero, we will use synthetic division to factor f.

$$\begin{array}{r|rrrr} -1 & 2 & -2 & -74 & -70 \\ & & -2 & 4 & 70 \\ \hline & 2 & -4 & -70 & 0 \end{array}$$

$\therefore f(x) = (x + 1)(2x^2 - 4x - 70) = 2(x + 1)(x^2 - 2x - 35) = 2(x + 1)(x - 7)(x + 5).$

30.4   Spring 2005 Final Solutions

1. The slope between the points $(7, 4)$ and $(-13, 4)$ is given by
   $m = (4 - 4)/(-13 - 7) = 0/-20 = 0$. Because the slope is 0, the line
   through the two points is horizontal and has the equation $y = 4$. Thus, the
   y−intercept is 4. Thus, $m = 0$ and $b = 4$ and $m + b = 0 + 4 = 4$.
   The correct answer is **E**.

2. Solving $2x - 3y = 6$ for y to put the equation in slope-intercept form to find
   the slope, we get that $y = 2/3x - 2$. The slope of the line perpendicular to
   this line is $m = -3/2$. Only **A** has a slope of $-3/2$.
   The correct answer is **A**.

3. The only restrictions on the domain in this problem come about from
   instances of the denominator being 0. So the domain should include all
   values of x in which the denominator is not 0. To determine this, we will
   first find out where the denominator is 0 by solving the equation $x^2 - 1 = 0$.
   $x^2 - 1 = 0 \Rightarrow (x - 1)(x + 1) = 0 \Rightarrow x = \pm 1$.
   Thus the domain should include all values of x *except* $x = \pm 1$.
   The correct answer is **C**.

4. $\dfrac{1}{3}\log 8 - 2\log 4 = \log 8^{1/3} - log4^2 = \log 2 - \log 16 =$
   $\log \dfrac{2}{16} = \log \dfrac{1}{8}$.

   The correct answer is **A**.

5. From the equation $x - y = 2$, we see that $x = y + 2$. Substituting this into
   the other equation, we see get
   $2(y + 2) - 3y = 1 \Rightarrow 2y + 4 - 3y = 1 \Rightarrow -y = -3 \Rightarrow y = 3$.
   Substituting $y = 3$ into the first equation yields $x - (3) = 2 \Rightarrow x = 5$.
   Thus, $x + y = (5) + (3) = 8$.
   The correct answer is **B**.

6. The solution of $|2x - 1| > 7$ is equivalent to the solution of $(2x - 1) > 7$ or $-(2x - 1) > 7$.
   $(2x - 1) > 7 \Rightarrow 2x > 8 \Rightarrow x > 4$.
   $-(2x - 1) > 7 \Rightarrow 2x - 1 < -7 \Rightarrow 2x < -6 \Rightarrow x < -3$.

   Thus, $x > 4$ or $x < -3$.
   The correct answer is **A**.

7. The graph of $y = 2f(x - 1) + 4$ is shifted in the following way from the graph of $y = f(x)$. The graph is vertically stretched by a factor of 2, moved right 1 unit and moved up 4 units. The effects of this on the point $(3, 7)$ will move the point to $(3, 14)$, then $(4, 14)$ and then to $(4, 18)$.
   The correct answer is **D**.

8. Combined, Floyd and Levi are moving toward each other at 45 mph, so the trip will take 2 hours. Floyd will travel 25 mph for 2 hours or 50 miles.
   The correct answer is **C**.

9. From the Conjugate zeros theorem, the polynomial must also have the conjugate of $x = -2 + 3i$ as a zero. Thus, $-2 - 3i$ is also a zero.
   The correct answer is **D**.

10. $\log_2(\log(x))) = -1 \Rightarrow \log(x) = 2^{-1} \Rightarrow \log(x) = \frac{1}{2} \Rightarrow x = 10^{1/2}$.
    The correct answer is **A**.

11. $\log_5(110) = \dfrac{\log(110)}{\log 5} \approx \dfrac{2.0414}{.69897} \approx 2.92$.
    The correct answer is **B**.

12. $(f \circ g)(2) = f(g(2))$. From the graph of $y = g(x)$, we see that $g(2) = 3$. Thus, $f(g(2)) = f(3)$. From the graph of $y = f(x)$, we see $f(3) = 3$.
    The correct answer is **B**.

13. From the graph, we see that the function is increasing on $(-\infty, -1)$, decreasing on $(-1, 2)$ and increasing on $(2, \infty)$.
    The correct answer is **B**.

14. Because the horizontal asymptote of the given graph is at $y = -1$, we know that $b = -1$. Further, the graph of $y = 2^x$ contains a point that passes through $(0, 1)$. If the graph had been only shifted down 1 unit, then it would have a point through $(0, 0)$. From the given graph, we see that instead it has a point though $(3, 0)$. Thus, $a = 3$.
    The correct answer is **B**.

15. Checking each point, we see that:
    Checking $(0, 0)$ we see that $(0) + (0) = 0 \leq 1$, but $2(0) - 3(0) = 0 \not\geq 2$, so A.) is not a solution.
    Checking $(1, 1)$ we see that $(1) + (1) = 2 \not\leq 1$ and $2(1) - 3(1) = -1 \not\geq 2$, so B.) is not a solution .
    Checking $(-10, 0)$ we see that $(-10) + (0) = -10 \leq 1$, but $2(-10) - 3(0) = -20 \not\geq 2$, so C.) is not a solution.
    Checking $(2, 5)$ we see that $(2) + (5) = 0 \not\leq 1$ and $2(2) - 3(5) = -13 \not\geq 2$, so D.) is not a solution.
    Checking $(1, -1)$ we see that $(1) + (-1) = 0 \leq 1$ and $2(1) - 3(-1) = 5 \geq 2$, so because BOTH inequalities are made true, $(1, -1)$ is a solution.
    The correct answer is **E**.

16. $\sqrt{2x + 1} + 1 = x \Rightarrow \sqrt{2x + 1} = x - 1 \Rightarrow$
    $2x + 1 = x^2 - 2x + 1 \Rightarrow x^2 - 4x = 0 \Rightarrow$
    $x = 0, x = 4$ are potential solutions.
    Checking x=0
    $\sqrt{2(0) + 1} + 1 = 2 \neq 0 \Rightarrow x = 0$ is not a solution.
    Checking x=4
    $\sqrt{2(4) + 1} + 1 = 4 = 4 \Rightarrow x = 4$ is a solution.
    The correct answer is $x = 4$.

17. $1000e^{(0.05t)} = 2,300,000 \Rightarrow e^{(0.05t)} = 2,300 \Rightarrow$
    $0.05t = \ln(2,300) \Rightarrow t = \ln(2,300)/.05 \Rightarrow$
    $t = 154.81$.

18. Clearly, $f(x) = a(x - 1)^2 + 3$ and $f(-2) = 8$ so,
    $f(-2) = a((-2) - 1)^2 + 3 = 8 \Rightarrow$
    $9a + 3 = 8 \Rightarrow$
    $9a = 5 \Rightarrow$
    $a = 5/9 \Rightarrow$
    $f(x) = \frac{5}{9}(x - 1)^2 + 3.$

19. (a) $\pm 2, \pm 1$.

    (b) To find a rational zero, we'll use the remainder theorem. We see that $f(1) = 6$, $f(-1) = -2$, $f(2) = 28$ and $f(-2) = 0$. Thus, $x = -2$ is a rational zero.

Alternatively, synthetic division can be used to show that $x = -2$ is a rational zero.

$$
\begin{array}{r|rrrr}
-2 & 1 & 4 & 3 & -2 \\
   &   & -2 & -4 & 2 \\
\hline
   & 1 & 2 & -1 & 0
\end{array}
$$

This synthetic division shows that $f(x) = (x+2)(x^2 + 2x - 1)$.

(c) $f(x) = (x+2)(x^2 + 2x - 1)$. See above.

(d) Using the quadratic formula, we see that
$$
x = \frac{-2 \pm \sqrt{2^2 - 4 \cdot 1 \cdot -1}}{2(1)} = \frac{-2 \pm \sqrt{8}}{2}
$$
$$
= \frac{-2 \pm 2\sqrt{2}}{2} = 2\left(\frac{-1 \pm \sqrt{2}}{2}\right) = -1 \pm \sqrt{2}.
$$

20. (a) $f(2) = 3(2)^2 - 5 = 3(4) - 5 = 12 - 5 = 7.$

(b) $f(a+1) = 3(a+1)^2 - 5 = 3(a^2+2a+1) - 5 = 3a^2+6a+3-5 = 3a^2+6a-2.$

(c) $\dfrac{3(x + h)^2 - 5 - (3x^2 - 5)}{h} = \dfrac{3(x^2 + 2xh + h^2) - 5 - 3x^2 + 5}{h}$

$\dfrac{3x^2 + 6xh + 3h^2 - 5 - 3x^2 + 5}{h} = \dfrac{6xh + 3h^2}{h}$

$\dfrac{h(6x + 3h)}{h} = 6x + 3h \text{ or } 3(2x + h).$

21. $7 \geq 8 - 2x > -3 \Rightarrow$

$-1 \geq -2x > -11 \Rightarrow$

$1/2 \leq x < 11/2 \Rightarrow$
$[1/2\,,\, 11/2).$

30.5   Fall 2005 Final Solutions

1. The x-intercept occurs when $y = 0$.
   Thus, $2x - 4(0) + 16 = 0 \Rightarrow 2x = -16 \Rightarrow x = -8$.
   Only one answer has $(-8, 0)$ as an x-intercept.
   The correct answer is **D**.

2. The standard form of a circle is given by $(x - h)^2 + (y - k)^2 = r^2$ where
   (h, k) represents the center and r represents the radius. Thus, the answer
   here should be $(x - 4)^2 + (y + 4)^2 = 3^2$.
   The correct answer is **A**.

3. The given line $2x + y = 5$ is equivalent to $y = -2x + 5$. So the line
   perpendicular to the given line has slope $m = \dfrac{1}{2}$. The point-slope form of
   the line through the point $(1, -2)$ with slope $m = \dfrac{1}{2}$ is $y + 2 = \dfrac{1}{2}(x - 1)$.
   The correct answer is **A**.

4. If you have traveled 50 km in 30 minutes, then you would travel 100 km in
   one hour. Put another way, your rate is $100 \ \dfrac{km}{hr}$. Using the formula
   $D = rt$, we see that $200 = 100t \Rightarrow t = 2$.
   The correct answer is **C**.

5. For $x^2 + kx + 9 = 0$ to have exactly one real solution, the discriminant
   $(b^2 - 4ac)$ must equal zero. Thus,
   $k^2 - 4(1)(9) = 0 \Rightarrow k^2 = 36 \Rightarrow k = \pm 6$.
   The correct answer is **B**.

6. $6 \le 2x + 10 \le 12 \Rightarrow -4 \le 2x \le 2 \Rightarrow$
   $-2 \le x \le 1$.
   The correct answer is **A**.

7. $|3 - 4x| \le 7 \Rightarrow$

$3 - 4x \le$	7 and	$-(3 - 4x) \le$	$7 \Rightarrow$
$-4x \le$	4 and	$3 - 4x \ge$	$-7 \Rightarrow$
$x \ge$	-1 and	$-4x \ge$	$-10 \Rightarrow$
$x \ge$	-1 and	$x \le$	$5/2 \Rightarrow$

   The correct answer is **C**.

8. To find the domain of g, note that the values under the radical must be
   nonnegative and the values in the denominator must be nonzero. Thus,
   $x + 5 \ge 0 \Rightarrow x \ge -5$ and $x^2 - 4 \ne 0 \Rightarrow (x + 2)(x - 2) \ne 0 \Rightarrow x \ne \pm 2$.
   Because all other values are allowed, the answer should represent values
   greater than or equal to -5, except for ± 2.
   The correct answer is **B**.

9. By inspection, we see that that graph is going down from left to right up until -3 and then from -1 on.
The correct answer is **E**.

10. $(g - f)(2) = g(2) - f(2)$. By inspection, we see that $g(2) = 3$ and $f(2) = 1$.
So, $(g - f)(2) = 3 - 1 = 2$.
The correct answer is **D**.

11. Noting that the graph has been shifted down 6 and right 3 would yield the function $y = |x - 3| - 6$. However, this graph would be 'too wide' for the correct graph and it would not go through the point $(0, 0)$. It is also necessary to recognize that the graph has be stretched vertically by a factor of 2. Thus, the correct function is $y = 2|x - 3| - 6$.

The correct answer is **B**.

12. $x^\pi = e \Rightarrow \log_x e = \pi$.
The correct answer is **E**.

13. $2\left[\log_3 x - \dfrac{1}{2}\log_3(x+2) - 4\log_3(x-5)\right] = 2\log_3 x - \log_3(x+2) - 8\log_3(x-5) =$
$\log_3 x^2 - (\log_3(x+2) + \log_3(x-5)^8) = \log_3 x^2 - \log_3(x+2)(x-5)^8 =$

$\log_3 \dfrac{x^2}{(x+2)(x-5)^8}.$
The correct answer is **E**.

14. Going through the possibilities,
$y = c \cdot a^x$ has a horizontal asymptote at $y = 0$ and is always increasing.
$y = c \cdot b^x$ has a horizontal asymptote at $y = 0$ and is always decreasing.
$y = c \cdot \log_a(x)$ has a vertical asymptote at $x = 0$ and is always increasing.
$y = c \cdot \log_b(x)$ has a vertical asymptote at $x = 0$ and is always decreasing.
$y = c \cdot x^a$ depends on the value of a. Some of the graphs will be U-shaped (e.g. $x = 2$ or $x = 4$), some will have the shape of a cubic graph (e.g. $x = 3$ or $x = 5$) and some will have a shape similar to that of $y = \sqrt{x}$ (e.g. $x = 3/2$). None of these is the graph given.
The correct answer is **B**.

15. The correct answer is **C**.

16. $7(4^x) = 15 \Rightarrow 4^x = \dfrac{15}{7} \Rightarrow$
$x \log(4) = \log \dfrac{15}{7} \Rightarrow x = \dfrac{\log \frac{15}{7}}{\log 4}.$
The correct answer is **E**.

17. (a) If $f(x) = |x|$, then we need to find the 'inside' function that gives $f(g(x)) = |3 + 5x^4|$. Here, this inside function should be $g(x) = 3 + 5x^4$. A quick check would verify this as correct.

    (b) If $g(x) = x^4$, then we need to determine what is being done with the x^4 to yield $3 + 5x^4$. $g(x)$ is being multiplied by 5, adding 3 and then the absolute value of all of this is being taken. Thus, $f(x) = |3 + 5x|$. Again, a check of $(f(g(x))$ verifies that this is correct.

18. (a) $f(-2) = 1$ since $x = -2$ corresponds to $x \le -2$.

    (b) $f(4) = 1 - \sqrt{4} = 1 - 2 = -1$ since $x = 4$ corresponds to $x > 1$.

    (c)

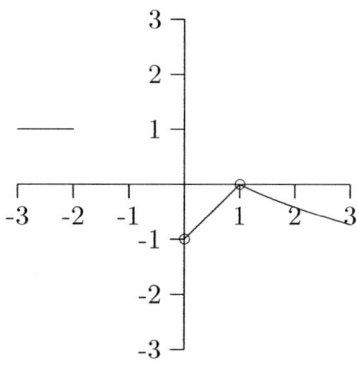

19. (a) $\pm 1, \pm 2, \pm 5, \pm 10$

2	1	-3	5	-1	-10
		2	-2	6	10
	1	-1	3	5	0

    $\therefore g(x) = (x - 2)(x^3 - x^2 + 3x + 5)$.
    For simplicity, let $f(x) = x^3 - x^2 + 3x + 5$. So, if $f(a) = 0$, then $g(a) = 0$.
    Using the possible rational zeros, we see that $f(1) = 8$, $f(-1) = 0$. So, $x = -1$ is a zero of g.

    (b) The results for synthetic division with f and $x = -1$ are below.

-1	1	-1	3	5
		-1	2	-5
	1	-2	5	0

    So, $g(x) = (x - 2)(x + 1)(x^2 - 2x + 5)$.

    (c) Because g has real coefficients, if $1 + 2i$ is a zero, then its conjugate $1 - 2i$ must be also. Thus,
    $g(x) = (x - 2)(x + 1)(x - (1 + 2i))(x - (1 - 2i)) = (x - 2)(x + 1)(x - 1 - 2i)(x - 1 + 2i)$.

20. Using the continuously compounded interest formula $A = Pe^{rt}$, we see in this problem that $A = 500$, $P = 300$ and $t = 4$. Thus we need to solve $500 = 300e^{r4}$ for r.

$$500 = 300e^{r4} \Rightarrow \frac{500}{300} = e^{4r} \Rightarrow$$

$$\ln\left(\frac{5}{3}\right) = 4r \Rightarrow r = \frac{\ln\left(\frac{5}{3}\right)}{4} \Rightarrow$$

$r \approx .1277$ or 12.77%.

21. First on this problem, note that the entire inequality lies in the first quadrant, so $x \geq 0, y \geq 0$.

One of the boundary lines contains the points $(7,0)$ and $(0,7)$. Using the slope formula, we see that $m = -1$ and the line has the equation $y = -x + 7$. Noting that the shaded area includes $(1,1)$ implies that the correct inequality must be such that $1 ? - (1) + 7$ is true when the ? is replaces with either $\geq$ or $\leq$. The $\leq$ makes the inequality true, so $y \leq -x + 7$ is another inequality that helps form the graph.

The remaining boundary line contains the points $(0,5)$ and $(1,6)$. Using the slope formula, we see that $m = 1$ and the line has the equation $y = x + 5$. Checking which inequality is made true with the test point $(1,1)$, we see that $y \geq x + 5$ is the correct inequality.

The complete system that yields the given graph is below.

$$\begin{cases} y \leq -x + 7 \\ y \geq x + 5 \\ x \geq 0 \\ y \geq 0 \end{cases}$$

30.6   Spring 2006 Final Solutions

1. The slope between the points is $m = \dfrac{4-1}{-3-2} = -\dfrac{3}{5}$. So the equation of the line in point-slope form is
$$y - 1 = -\frac{3}{5}(x-2) \Rightarrow 5y - 5 = -3x + 6 \Rightarrow 3x + 5y - 11 = 0.$$
The correct answer is **E**.

2. Since $x = -7$ is a vertical line, the line perpendicular to it must be horizontal. The horizontal line through $(-6, 1)$ is $y = 1$.
The correct answer is **A**.

3. $\sqrt{-10}\sqrt{-5} = i\sqrt{10}\,i\sqrt{5}$.
The correct answer is **D**.

4. $x^2 + 5x = 10 \Rightarrow x^2 + 5x - 10 = 0$. Using the quadratic formula yields
$$x = \frac{-5 \pm \sqrt{5^2 - 4(1)(-10)}}{2(1)} = \frac{-5 \pm \sqrt{65}}{2}.$$
The correct answer is **D**.

5. $\dfrac{1}{x-2} \le \dfrac{1}{x^2-4} \Rightarrow \dfrac{1}{x-2} - \dfrac{1}{x^2-4} \le 0 \Rightarrow$
$\dfrac{x+2}{x^2-4} - \dfrac{1}{x^2-4} \le 0 \Rightarrow \dfrac{x+1}{(x+2)(x-2)} \le 0.$

Intervals	$(-\infty, -2)$	$(-2, -1]$	$[-1, 2)$	$(2, \infty)$
Test points	$x = -3$	$x = -1.5$	$x = 0$	$x = 3$
$\dfrac{x+1}{(x+2)(x-2)}$	$-\dfrac{2}{5}$	$\dfrac{2}{7}$	$-\dfrac{1}{4}$	$\dfrac{4}{5}$
Signs of $\dfrac{x+1}{(x+2)(x-2)}$	- negative	+ positive	- negative	- positive

The correct answer is **A**.

6. $f(a+h) - f(a) = (a+h)^2 - 4 - (a^2 - 4) = a^2 + 2ah + h^2 - 4 - a^2 + 4 = 2ah + h^2.$
The correct answer is **B**.

7. $a^c = b \Rightarrow \log_a(b) = c.$
The correct answer is **A**.

8. The possible rational zeros of P are the factors of 9 divided by the factors of 24. Since 2 is not a factor of 9, $\dfrac{2}{3}$ is not a possible rational zero of P.
The correct answer is **E**.

9. The graph is reflected across the x-axis, shifted right 3, up 1 and is vertically stretched by a factor of 2. So the correct graph is
$$y = -2(x-3)^2 + 1.$$
The correct answer is **B**.

10. Logarithms require positive inputs. Thus, $2x + 7 > 0 \Rightarrow x > -7/2$.
The correct answer is **C**.

11. $(fg)(1) = f(1)g(1)$. By inspection, we see that $f(1) = -3$ and $g(1) = -1$, so $(fg)(1) = (-3)(-1) = 3$.
    The correct answer is **B**.

12. $\ln(x) - \ln(x - 4) = 5 \Rightarrow \ln\left(\dfrac{x}{x-4}\right) = 5 \Rightarrow$

    $\dfrac{x}{x-4} = e^5 \Rightarrow x = (x-4)e^5$

    $x = e^5 x - 4e^5 \Rightarrow 4e^5 = e^5 x - x \Rightarrow$

    $4e^5 = x(e^5 - 1) \Rightarrow x = \dfrac{4e^5}{e^5 - 1}$.

    The correct answer is **B**.

13. The leading term $-x^5$ determines the end behavior.
    The correct answer is **C**.

14. According to the remainder theorem, $f(-2) = 0$. So,
    $f(-2) = (-2)^3 - 4(-2)^2 + k(-2) - 6 = 0 \Rightarrow -8 - 16 - 2k - 6 = 0 \Rightarrow$
    $-2k = 30 \Rightarrow k = -15$.
    The correct answer is **B**.

15. First on this problem, note that the entire inequality lies in the first quadrant, so $x \geq 0, y \geq 0$.
    One of the boundary lines contains the points $(5, 0)$ and $(0, 10)$. Using the slope formula, we see that $m = -2$ and the line has the equation $y = -2x + 10$. Noting that the shaded area includes $(1, 4)$ implies that the correct inequality must be such that $4? - 2(1) + 10$ is true when the ? is replaces with either $\geq$ or $\leq$. The $\leq$ makes the inequality true, so $y \leq -2x + 10$ is another inequality that helps form the graph.
    The remaining boundary line contains the points $(0, 1)$ and $(4, 2)$. Using the slope formula, we see that $m = \dfrac{1}{4}$ and the line has the equation $y = \dfrac{1}{4}x + 1$. Checking which inequality is made true with the test point $(4, 1)$, we see that $y \geq \dfrac{1}{4}x + 1$ is the correct inequality.
    The correct answer is **C**.

16. Let x = amount on necessary fruit juice concentrate necessary, in liters. Then,
    $x \cdot .8 = 15 \cdot .2 \Rightarrow$
    $x = 3/.8 = 3.75$ liters.

17. $1000000 = 12000 \left(\dfrac{1.09^n - 1}{.09} \right) \Rightarrow \dfrac{1000000}{12000} = \dfrac{1.09^n - 1}{.09} \Rightarrow$

    $.09(\dfrac{1000000}{12000}) = 1.09^n - 1 \Rightarrow .09(\dfrac{1000000}{12000}) + 1 = 1.09^n \Rightarrow$

    $\log(.09(\dfrac{1000000}{12000}) + 1) = n \log(1.09) \Rightarrow$

    $n = \dfrac{\log(.09(\frac{1000000}{12000}) + 1)}{\log(1.09)} \approx 24.83 = 25$ years.

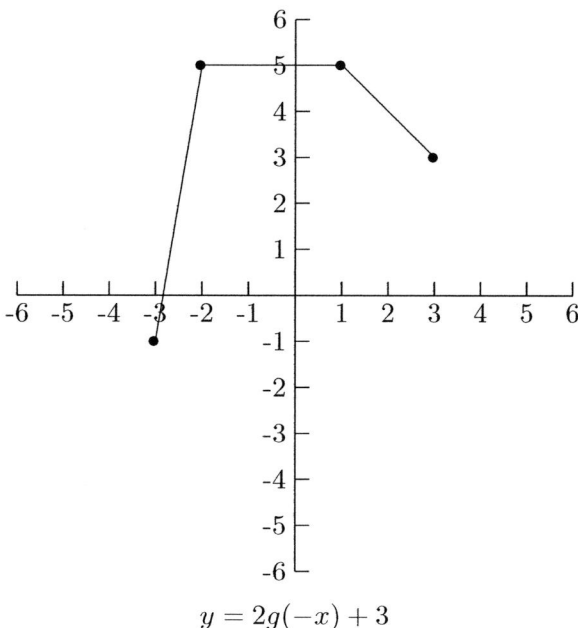

$y = 2g(-x) + 3$

18.

19. (a) $f(-2) = 1$.

     (b) $f(0) = 3$.

     (c) $f(f(5)) = 2$.

     (d) Sketch the graph of $y = f(x)$ on the axes provided below.

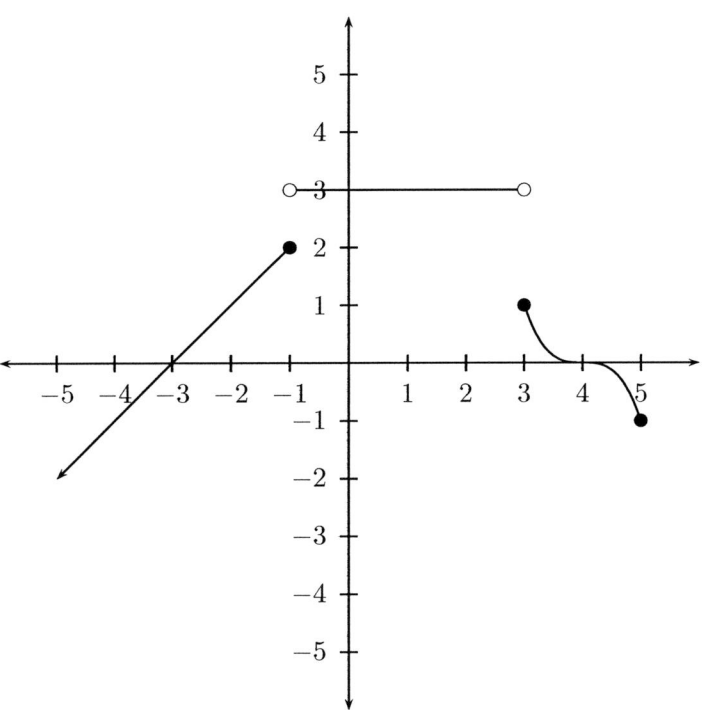

20. $-|3x - 7| + 10 < 8 \Rightarrow$
     $-|3x - 7| < -8 \Rightarrow |3x - 7| > 8 \Rightarrow$
     $3x - 7 > 8$ OR $-(3x - 7) > 8 \Rightarrow$
     $3x > 15$ OR $(3x - 7) < -8 \Rightarrow$
     $x > 5$ OR $3x < -1 \Rightarrow$
     $x > 5$ OR $x < -1/3 \Rightarrow$
     $(-\infty, -1/3) \cup (5, \infty)$

21. The solution using elimination is shown below:
$$2x + 4y = -4$$
$$5x - 7y = 41 \Rightarrow$$
$$14x + 28y = -28$$
$$20x - 28y = 164 \Rightarrow$$
$$34x = 136 \Rightarrow x = 4$$
$$2x + 4y = -4 \Rightarrow 2(4) + 4y = -4 \Rightarrow 4y = -12 \Rightarrow y = -3.$$

22. (a)

$$
\begin{array}{r|rrrr}
-1 & 3 & -10 & 1 & 14 \\
  &   & -3 & 13 & -14 \\
\hline
  & 3 & -13 & 14 & 0
\end{array}
$$

(b) Use the remainder theorem to show that $x - 2$ is a factor of $f(x)$.
$f(2) = 3(2)^3 - 10(2)^2 + (2) + 14 = 24 - 40 + 2 + 14 = 0$. Thus 2 is a root of f and $x - 2$ is a factor.

(c) Completely factor $f(x)$.
$$f(x) = (x + 1)(3x^2 - 13x + 14) = (x + 1)(x - 2)(3x - 7).$$

Index